F-22A Multiyear Procurement Program

An Assessment of Cost Savings

Obaid Younossi • Mark V. Arena • Kevin Brancato • John C. Graser
Benjamin W. Goldsmith • Mark A. Lorell • Fred Timson • Jerry M. Sollinger

Prepared for the Office of the Secretary of Defense

The research described in this report was prepared for the Office of the Secretary of Defense (OSD). The research was conducted in the RAND National Defense Research Institute, a federally funded research and development center sponsored by the OSD, the Joint Staff, the Unified Combatant Commands, the Department of the Navy, the Marine Corps, the defense agencies, and the defense Intelligence Community under Contract W74V8H-06-0002.

Library of Congress Cataloging-in-Publication Data

F–22A multiyear procurement program : an assessment of cost savings / Obaid Younossi ... [et al.].
p. cm.
Includes bibliographical references.
ISBN-13: 978-0-8330-4196-8 (pbk. : alk. paper)
1. United States. Air Force—Procurement—Evaluation. 2. F/A–22 (Jet fighter plane)—Costs. I. Younossi, Obaid. II. Title: F–22 multiyear procurement program.

UG1123.F15 2007
358.4'16212—dc22

2007031096

Cover Photo Provided by the F-22A System Program Office

The RAND Corporation is a nonprofit research organization providing objective analysis and effective solutions that address the challenges facing the public and private sectors around the world. RAND's publications do not necessarily reflect the opinions of its research clients and sponsors.

Cover Design by Stephen Bloodsworth

Published 2007 by the RAND Corporation
1776 Main Street, P.O. Box 2138, Santa Monica, CA 90407-2138
1200 South Hayes Street, Arlington, VA 22202-5050
4570 Fifth Avenue, Suite 600, Pittsburgh, PA 15213-2665
RAND URL: http://www.rand.org/
To order RAND documents or to obtain additional information, contact
Distribution Services: Telephone: (310) 451-7002;
Fax: (310) 451-6915; Email: order@rand.org

Preface

This document represents the response to a requirement by the Congress of the United States in the Fiscal Year (FY) 2007 National Defense Authorization Act for an independent review, assessment, and analysis of the savings related to using a multiyear contract to buy 60 F-22A aircraft and associated engines during FYs 2007 through 2009. The U.S. Air Force was in the process of awarding multiyear contracts to the Lockheed Martin–Boeing team and Pratt & Whitney, which are the prime contractors for the F-22A.[1] Congress wanted assurance that the projected savings claimed for the multiyear contract in 2006 were feasible and specifically asked for a comparison between the multiyear contracts and single-year contracts that would buy 20 aircraft and associated engines each year for three years. Moreover, Congress asked for a consideration of historical cost savings on other aviation-related multiyear contracts dating back to 1982.[2]

The RAND Corporation's National Defense Research Institute was asked by the Office of the Secretary of Defense to carry out such a study. The study was conducted from November 2006 through June 2007. This report should interest those involved in the acquisition of weapons systems and those concerned with assessing the cost of such systems.

---

[1] The U.S. Air Force awarded the contracts on July 31, 2007.

[2] The results of this study were briefed to the Office of the Under Secretary of Defense for Acquisition, Technology and Logistics on June 21, 2007, and a pre-publication version of the report was delivered to Congress on June 29, 2007.

The authors welcome your questions and comments regarding this research, which can be addressed to

Obaid Younossi
RAND Corporation
1200 South Hayes Street
Arlington, VA 22202-5050
Tel: 703-413-1100, x5235
Email: Obaid_Younossi@rand.org

National Defense Research Institute (NDRI)

This research was sponsored by the Office of the Under Secretary of Defense for Acquisition, Technology and Logistics and conducted within the Acquisition and Technology Policy Center of the RAND National Defense Research Institute, a federally funded research and development center sponsored by the Office of the Secretary of Defense, the Joint Staff, the Unified Combatant Commands, the Department of the Navy, the Marine Corps, the defense agencies, and the defense Intelligence Community.

For more information on RAND's Acquisition and Technology Policy Center, contact the Director, Philip Antón. He can be reached by email at atpc-director@rand.org; by phone at 310-393-0411, extension 7798, or by mail at the RAND Corporation, 1776 Main Street, Santa Monica, California 90407-2138. More information about RAND is available at http://www.rand.org.

Dedication

We dedicate this report to our colleague, friend, and mentor, Bob Roll. We will miss his guiding hand.

Contents

Figures

Tables

Summary

Acquiring defense weapon systems under multiyear contracts rather than a series of single-year contracts offers the possibility of cost savings. Such contracts afford contractors the opportunity to buy materials in more economical quantities, schedule workers and facilities more efficiently, and reduce the burden of preparing multiple proposals. The U.S. government also benefits from a reduced workload. The U.S. Air Force is in the process of awarding a multiyear contract for 60 F-22A aircraft over three years. Congress wants to assure itself that the proposed contract will yield the promised savings and asked, in the FY 2007 National Defense Authorization Act, for an independent review of the estimated savings. RAND's National Defense Research Institute was asked to conduct the review by the Office of the Under Secretary of Defense for Acquisition, Technology and Logistics.

Purpose and Approach

The National Defense Authorization Act asked for two things. First, it requested a comparison of the multiyear procurement of 60 F-22A aircraft and associated engines with three single-year procurements of 20 aircraft and engines. Second, it asked for a comparison between historical cost savings achieved for aviation-related multiyear contracts dating back to fiscal year 1982 and the projected savings of the F-22A multiyear contract.

To comply with this congressional request, the research team identified three tasks:

1. Estimate the costs for baseline procurement of 60 aircraft and associated engines, including spare engines, in FYs 2007–2009 under annual (single-year) contracts at the rate of 20 aircraft per year.
2. Substantiate contractor-proposed savings, and compare them to the difference between multiyear negotiated prices and RAND's single-year price estimates.
3. Report the cost savings resulting from historical and ongoing aviation-related (aircraft and aircraft engines) multiyear procurement contracts back to FY 1982.

To accomplish these tasks, the research team reviewed existing literature on multiyear contracts in other aircraft programs and analyzed historical data. Team members also reviewed both the multiyear proposals for Lots 7, 8, and 9 and the Lot 7 single-year proposals submitted by the F-22A air vehicle and engine prime contractors. They visited and collected considerable information and historical data from the F-22A System Program Office and prime contractors—Lockheed Martin (LM), Boeing, and Pratt & Whitney (P&W)—and some of the major subcontractors involved in the production of the F-22A. They also visited and gathered information from the C-17, C-130, and F/A-18E/F program offices, as well as the Institute for Defense Analyses (IDA).

Results

The results of our assessment of the single-year procurement, our substantiation of contractor savings, and our analysis of historical savings appear below. Although some historical comparisons in this report are shown in base year dollars, we primarily present our results in then-year (TY) dollars.[3]

[3] We use this format for four reasons: (1) Wide discrepancies exist between various indices for deflating TY to BY. These discrepancies mask the actual contract prices and savings that were negotiated or estimated. (2) Congressional interest is in annual budgets, which are authorized and appropriated in then-year dollars. (3) The negotiated prices between the

Cost of Single-Year Procurement (SYP)

One of the more challenging tasks in assessing the realism of cost savings due to a multiyear procurement strategy is evaluating the costs of the alternative single-year contract strategy (the path not taken). We estimated the three single-year equivalent prices for FYs 2007–2009 using the same program content as that of the multiyear proposal, along with F-22A historical cost data, negotiated proposal values, and three alternative cost improvement curve (CIC)[4] assumptions, as follows:

A. The downward trend in costs experienced during Lots 1–6 will continue through Lots 7–9. Thus, costs for Lots 7–9 would reflect the same cost improvement curve.
B. The trend for Lots 5 and 6 is more indicative of what will occur in Lots 7–9. In this assumption, a new CIC was developed using the same cumulative quantities and lot midpoints as above, except that the Lot 5 and 6 costs were used in the regression. The new CIC was used to predict Lot 7–9 costs. These resulting costs were adjusted for the change in annual production rate from Lot 6 to Lot 7 and beyond.
C. Lot 6 cost data are the most indicative of likely costs for the next three lots. Lot 6 costs (adjusted for annual rate of inflation) were used for Lots 7–9.

Added to the above three assumptions was the cost of the engines, both those to be installed at the end of the air vehicle production (120 engines) and spare engines (13) to be procured under multiyear procurement (MYP). Propulsion costs were based on unit price data for

USAF and contractors for both the single-year and multiyear contracts and for the savings initiatives are all in TY dollars. By leaving the contractor proposals in TY dollars, we avoid distorting the numbers. (4) Formal justifications for historical multiyear aircraft procurements were part of the DoD budgeting and congressional justification processes and were shown only in TY dollars.

[4] A *cost improvement curve* is similar to a learning curve (wherein average labor hours per lot decrease at a certain percentage as cumulative quantities double), except that recurring costs are substituted for labor hours in the CIC equation. (Younossi, Kennedy, and Graser, 2001).

whole engines and were not broken out by labor hours and material because of the limited data available.

The differences between our single-year estimate and the negotiated multiyear proposals are shown in Table S.1.

Based on our model, we estimate the savings range to be between $274 million and $643 million and our most realistic estimate to be $411 million.[5]

Substantiation of Contractor-Proposed Savings

To better assess the realism of our savings estimate, we also undertook to substantiate the MYP savings proposed by the contractors. As part of the pre-award activities for the multiyear contract for Lots 7–9, the U.S. Air Force (USAF) asked Boeing, Lockheed Martin, Pratt & Whitney, and their subcontractors and suppliers to develop cost savings initiatives that could be implemented as part of the contract. We substantiated the reasonableness of the proposed savings associated with these initiatives as follows: We reviewed their methodology for computing the savings by comparing the single-year cost estimate with the multiyear cost for each initiative, evaluated the feasibility of the initiative, and ensured that the savings were incorporated into the negotiated values for the multiyear contract by tracing each initiative into the final con-

Table S.1
Estimated Multiyear Savings CIC Assumption (TY $millions)

CIC Assumption	Lot 7 Savings	Lot 8 Savings	Lot 9 Savings	Total Savings
Lots 1–6	43	108	123	274
Lots 5–6	48	153	210	411
Lot 6	65	229	349	643

[5] These estimates do not include any potential savings due to reduced program office workload from not having to write yearly contracts. Estimating those further savings was outside the scope of this study, and historically they are not included in the estimates of MYP savings.

tract price.[6] We organized savings into six categories. We also noted aspects of the F-22A multiyear contract that could not be quantified in dollar terms per se but that are positive aspects of a multiyear contract. In addition, we noted that certain activities remained basically the same under either case because the overall program is not being accelerated to a more efficient annual rate due to the constrained budget available. The total value of the savings proposed by the contractors was $311 million in TY dollars; we substantiated $296 million. Almost three-fourths of the savings were accounted for in two categories: the buy-out of materials and parts and support labor savings. The $296 million in substantiated savings for the contractor initiatives falls within the range of our estimated savings of $274 million to $643 million using the alternative CIC analysis (see Table S.1). Therefore, we found that about 70 percent of our most realistic estimate of $411 million can be traced to substantiated contractors' identified savings estimates.

Historical Savings

Our analysis of historical fixed-wing aircraft multiyear procurement showed that program savings estimates from 1982 through 2007 varied from 5.5 percent to 17.7 percent. Estimated savings for fighter aircraft multiyear programs during this period ranged from 5.7 percent to 11.9 percent. For multiyear programs after 1995, we could find no significant statistical correlation between the contract savings estimates and such factors as contract size, total number of aircraft procured, number of aircraft procured annually, length of the MYP program, or funding provided for economic order quantity (EOQ) or cost reduction

---

[6] As part of the contract award process, the USAF asked the prime contractors to submit two proposals each. One addressed a single-year contract for 20 aircraft (Lot 7) in FY 2007 and the other addressed a multiyear contract for 60 aircraft (Lots 7–9) in FYs 2007–2009. Since the multiyear award was contingent upon a certification by the Secretary of Defense, both the single-year and multiyear options were fully negotiated and prepared for award, because neither the USAF nor the contractors knew whether the multiyear option would be approved. These negotiated proposals included negotiated prices between the prime contractors and their subcontractors and suppliers. As a result, the negotiated single-year Lot 7 values were a source of valuable data in establishing savings between the two options.

initiatives.[7] However, we note that this does not necessarily mean that no correlations exist among these or other factors and the magnitude of estimated savings. The problem is that we do not have enough data points to support a statistically valid correlation analysis. Correlations may exist, but they cannot be statistically demonstrated with this limited database.

Qualitative examination of the post-1995 multiyear programs and many of the pre-1995 programs suggested that achieving subcontractor and vendor quantity discounts is a key factor in obtaining savings on multiyear procurement programs. Achieving significant savings in this area, however, did not always appear to hinge on EOQ funding. Moreover, our qualitative assessment of the historical programs' characteristics that contributed to the multiyear cost savings—such as total number and rate of aircraft procured under the multiyear program, EOQ funding and timing of the funding, timing of the multiyear contract implementation, and availability of funding for cost reduction initiatives—indicates that these characteristic were more favorable to the historical programs than to the F-22A multiyear program because the F-22A MYP differs in many key characteristics from historical multiyear programs.

F-22A MYP Savings

We calculated the savings percentage by dividing our estimate of the single-year contract for Lots 7, 8, and 9 by the difference between our SYP estimates and the MYP negotiated prices. Table S.2 shows our percentage of estimated savings.

The substantiated contractor savings that would result from the F-22A firm-fixed-price multiyear contracts for 60 aircraft, instead of three separate single-year contracts for lot sizes of 20 aircraft each, is about $296 million (TY), or 3.3 percent, for the three single-year contracts. This percentage falls within the range of our estimated savings of 3.1 percent, in our most pessimistic case, and 6.9 percent, in our most optimistic case.

---

[7] *Economic order quantity* refers to the optimal quantity to order that minimizes total variable costs required to order and hold inventory.

Table S.2
RAND's Estimated Savings Percentages

CIC Assumption	Single-Year Estimate (TY $millions)	Savings, SYP Minus MYP (TY $millions)	Savings Percentage
Lots 1 to 6	8,952	274	3.1%
Lots 5 and 6	9,089	411	4.5%
Lot 6	9,320	643	6.9%

Figures S.1 and S.2 provide two additional ways of thinking about the savings for the multiyear procurement by comparing the F-22A estimated savings with those of the savings estimated for historical programs.[8] The first arrays the results of our research as a function of the percentage of contract savings relative to a single-year procurement contract. The leftmost bar shows historical data for all fixed-wing aircraft that underwent a multiyear procurement from 1982 to 2007 and indicates the low, median, and high values. The next bar presents the data for fighter and attack aircraft only, displayed in a similar fashion. Finally, the last bar shows savings based on assumptions about SYP learning curves (SYP1 reflects our first assumption, SYP2 reflects our second assumption, and SYP3 reflects our third assumption). Our estimated savings are low by estimated historical values both for the all-aircraft programs mix and for just fighter and attack aircraft.

Another way to evaluate savings is by the dollars saved per aircraft. Figure S.2 is similar to Figure S.1, except that the metric is dollar savings per aircraft in FY 2005 dollars. In this case, our estimated savings are high by historical standards of fighter and attack aircraft but well within the historical range of all aircraft MYP since 1982, which can be partially explained by the higher unit cost of the F-22A compared with other fighters.

---

[8] In general, the savings shown for historical programs were *estimates of savings* made as part of the multiyear contract budget justifications (pre-contract award) and are not historical "actual" savings per se.

Figure S.1
Savings Percentage Relative to SYP Value

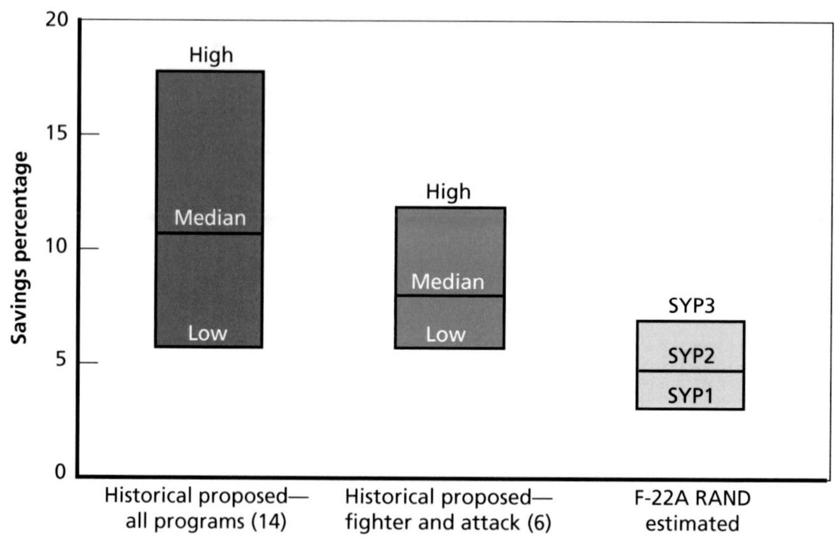

NOTE: The numbers in parentheses in the first two bar labels show the number of aircraft programs included in the analysis.

RAND *MG664-S.1*

Overall Findings

We find that examining the issue of multiyear savings using several approaches produces a consistent range of results.

- *We estimate cost savings due to the F-22A MYP contract to be about $411 million compared with three separate SYP contracts.* Estimating the three single-year contract prices and comparing them with the recently negotiated multiyear contracts for air vehicles and engines between the USAF and the F-22A contractors produces a range of estimates. When using the second cost improvement assumption for predicting these costs, we found that our prediction and the separate, negotiated single-year Lot 7 price were similar. This method resulted in overall savings of $411 mil-

Figure S.2
Dollar Savings per Aircraft

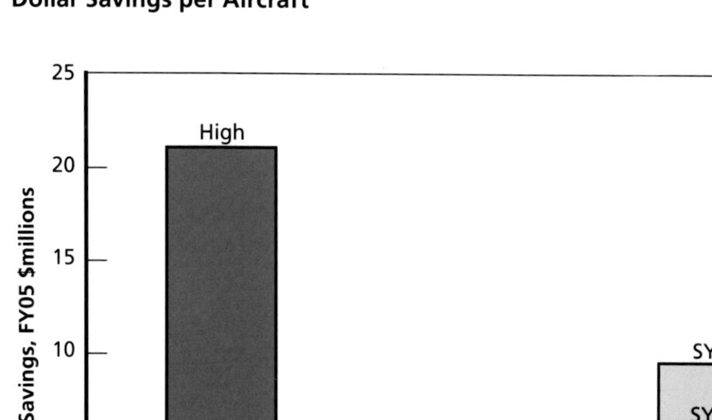

NOTE: The numbers included in the parentheses in the first two bar labels show the number of aircraft included in the analysis.

RAND *MG664-S.2*

lion, which translates to about 4.5 percent of the overall contract value.

- *More than 70 percent of the savings can be supported by substantiated contractors' savings initiatives.* Our substantiation of the contractor-proposed savings initiatives produced savings estimates of $296 million. We substantiated these savings by tracing them to the final negotiated MYP contract.
- *F-22A MYP contract savings percentages are lower than the historical range, but the program has fewer cost savings opportunities than the historical programs we analyzed.* Although the savings percentage for the multiyear prices compared with the single-year predicted prices is relatively low by historical estimated standards (4.5 percent), the dollar savings per aircraft in the F-22A multiyear contract are on the upper end of the savings of previous fighter/attack multiyear contracts (a result of the higher unit cost for the F-22A). In addition, our qualitative assessment of the pro-

gram attributes that contributed to the cost savings on historical multiyear programs indicates that the F-22A MYP differs in many key characteristics from historical MYPs, often in ways that put the F-22A MYP program at a disadvantage when compared with the historical MYPs.

Acknowledgments

The research and analysis presented in this report would not have been possible without the support of the Office of the Under Secretary of Defense for Acquisition, Technology and Logistics; the U.S. Air Force F-22A Program Executive Office (PEO) and F-22A System Program Office (SPO); Lockheed Martin and Boeing, the F-22A air vehicle prime contractors; their major subcontractors; and Pratt & Whitney (P&W), the manufacturer of the government-furnished-equipment F119 engine.

More specifically, the authors are grateful to David Hersh, OSD/AT&L, for sponsoring this project; Maj Gen Jeffery Riemer, F-22A PEO, for supporting the work and opening doors at the F-22A SPO and contractors; Col Michael Braman, the F-22A MYP contract manager, for coordinating the data collection visits, meetings, and access to his very busy team; and Doug Mangen and Ken Birkofer, from the F-22A SPO, for providing F-22A historical cost information and much more. We would like to thank Joseph Cardarelli, Mark Mutschler, and Doug Harris of Naval Air Systems Command for all their help. Our IDA colleagues Bruce Harmon, Richard Nelson, and John Hiller met with us and shared lessons learned, and for that we are grateful.

We also would like to thank Jack Twedell, Larry Pierce, Mark Byars, and Nick Cessario from Lockheed Martin; Bill Cribb, Kevin Flannigan, and Bob Jenkins from Boeing; and Chris Flynn, Jeff Zotti, and Trish Walker from P&W for hosting meetings at their facilities, providing data, and visiting RAND on a number of occasions. Without their and their team's data and insights, this study would not have been possible.

We are also grateful to our RAND colleagues Laura Baldwin and Frank Camm for carefully reviewing the draft manuscript and suggesting many substantive changes that enormously helped the readability and the quality of this report. We thank Brian Grady for research and administrative support and Miriam Polon for editing the report.

Finally, other RAND colleagues who provided helpful insights and encouragement during the study are Gene Gritton, Andy Hoehn, Phil Antón, Sue Bodily, Cynthia Cook, and Natalie Crawford. Lastly, we acknowledge the many, many contributions of our late friend and colleague C. Robert Roll, who was involved in the concept formulation of this project.

Abbreviations

AME	alternative mission equipment
AP	advance procurement
APN	Aircraft Procurement, Navy
ATF	Advanced Tactical Fighter
AWACS	Airborne Warning and Control System
BCA	business case analysis
BLS	Bureau of Labor Statistics
BTL	below the line
BY	base year
CBA	cost benefit analysis or common backplane assembly
CCDR	Contractor Cost Data Reporting
CDE	commercial derivative engine
CDSR	Cost Data Summary Reports
CFE	contactor furnished equipment
CIC	cost improvement curve
CLIN	contract line item number
CPAF	cost plus award fee
CRI	cost reduction initiative
CSDRP	Cost and Software Data Report
CSE	common support equipment
CY	current year or calendar year
DAPP	defense acquisition pilot program
DCMC	Defense Contract Management Command
DFMA	design for manufacturing and assembly
DMS	diminished manufacturing sources
DoD	Department of Defense
E&IS	electrical and intelligence systems
ECP	engineering change proposal

EMD	engineering and manufacturing development
EOQ	economic order quantity
EPA	economic price adjustment
EPIP	electronic product improvement process
EW	electronic warfare
FAR	Federal Acquisition Regulation
FCHR	functional cost hours reports
FFP	Firm fixed price
FPI	fixed price incentive
FPRA	forward priced rate agreement
FS&T	field support and training
FY	fiscal year
GFE	government furnished equipment
GFP	government furnished property
HAPCA	Historical Aircraft Procurement Cost Archive
HDMI	High Density Multilayer Interconnect
IDA	Institute for Defense Analyses
IDS	integrated defense solution
IFF	identification, friend or foe
IMIS	integrated maintenance information systems
IOC	initial operating capability
JCM	joint cost model
JSF	joint strike fighter
JTIDS	Joint Tactical Information Distribution System
LC	learning curve
LM	Lockheed Martin
LRIP	low-rate initial production
MILCON	military construction
MLRS	Marine Corps Long Range Study
MSS	mission support systems
MYP	multiyear procurement
NAVAIR	Naval Air Systems Command
NDAA	non-development aircraft acquisition
NDI	non-development item
NEAT	Nacelle Engine Affordability Team
O&M	operations and maintenance
OGC	other government cost
OPC	other production cost
OSD	Office of the Secretary of Defense

P&W	Pratt & Whitney
PALS	program agile logistics support
PB	President's Budget
PBA	price-based acquisition
PBD	program budget decision
PDS	passive defense system
PIP	productivity improvement program
PPBS	Programming, Planning and Budgeting System
PRTV	production representative test vehicles
PSAS	program support annual, sustaining
PSE	peculiar support equipment
PSO	product support, other
PSP	program support production
PSS	program support sustainment
PY	prior year
REDI	Raptor Enhanced Development and Integration
RFI	request for information
RFP	request for proposal
ROI	return on investment
SAR	Selected Acquisition Report
SBIRS	Space-Based Infrared System
SPO	System Program Office
SYP	single-year procurement
TINA	Truth in Negotiation Act
TPC	Total performance curve
TY	then year
UID	unique item identifier
USAF	United States Air Force
USC	United States Code
WBS	work breakdown schedule

CHAPTER ONE

Introduction

Background

This section first gives a brief history of the F-22A program and follows with an explanation of the congressional directive to the Department of Defense (DoD).

History of the F-22A Program

In November 1981, DoD formally launched the Advanced Tactical Fighter (ATF) program.[1] The ATF was intended to replace the F-15, then the Air Force's premier air superiority fighter. The U.S. aerospace industry realized that the ATF would be the only opportunity to develop an all-new, cutting-edge-technology supersonic fighter for the next decade or longer.

During 1982, a consensus began to emerge that a modified version of the F-15 or F-16 could perform the air-to-ground role, permitting the ATF to be optimized for air superiority. By mid-1983, the ATF had clearly been defined as the replacement for the F-15 air superiority fighter. Following the emergence of this consensus, the Air Force awarded concept development contracts in September 1983 for further refinement of the design concepts for the ATF.

The Air Force sent out requests for proposal (RFPs) for the demonstration/validation phase of the ATF in October 1985. Two

[1] The ATF program was the forerunner of the F-22 program, which later became the F/A-22 program. The Defense Resources Board first approved ATF program start-up on November 23, 1981.

teams were selected: one comprising Lockheed Martin, Boeing, and General Dynamics and the other comprising McDonnell Douglas and Northrop. Similarly, two engine alternatives were considered in each design, and each contractor team was required to demonstrate its airframe design with both the Pratt & Whitney F119 engine and General Electric's F120. In August, the Air Force awarded separate cost plus award fee (CPAF) engineering and manufacturing development (EMD) contracts to the Lockheed Martin–Boeing team for the air vehicle and support development and to Pratt & Whitney to develop the F119 engine and support. The engines are provided as government furnished equipment (GFE) to Lockheed Martin for installation.

With the end of the Cold War, the F-22A program experienced significant program perturbations. The original planned procurement quantity was successively reduced from 750 aircraft to the current level of 183, the two-seat variant was eliminated, and the aircraft was redesignated the F/A-22 to reflect the addition of its air-to-ground role. Its designation was later changed back to the F-22A.[2] At present, the Air Force states a need for 381 aircraft.

Congressional Directive to the Office of the Under Secretary of Defense for Acquisition, Technology and Logistics

In 2006, the Defense Appropriation Conference Report directed the Secretary of Defense to report on alternative procurement strategies for the F-22A program. This analysis is required by Title 10, USC 2306b, to justify that "substantial savings" would occur from the use of the multiyear contract. (The requirements of this law are discussed in the next chapter.) DoD then tasked the Institute for Defense Analyses (IDA) to assess potential savings. The IDA assessment, completed in May 2006, estimated savings at $235 million, or about 2.6 percent for air vehicle and 2.7 percent for engines, of the estimated total procurement cost of $8.7 billion for the 60 aircraft and 120 engines.[3] Appendix D provides a summary the IDA report.

---

[2] Younossi, Stem, et al., 2005.

[3] Nelson et al., 2006.

During congressional hearings on the fiscal year (FY) 2007 President's Budget (PB) submission, the Air Force proposed a change to the F-22A program in which the last 60 aircraft plus 120 installed and 13 spare engines would be procured under a multiyear contracting arrangement for FYs 2007 through 2009. The proposal was authorized, and funds were appropriated as part of the FY 2007 USAF procurement budget. However, Congress directed that another independent assessment of the multiyear savings be conducted. In November 2006, the Office of the Under Secretary of Defense for Acquisition, Technology and Logistics asked RAND's National Defense Research Institute (NDRI) for such an assessment.

Purpose of This Report

This report provides an independent assessment and analysis of the savings related to using a multiyear contract for the procurement of 60 aircraft and associated engines during FYs 2007–2009. This analysis is part of a certification by the Secretary of Defense required by the FY 2007 National Defense Authorization Act, Section 134. This certification must be completed 30 days before the award of the F-22A production contracts for FY 2007.

Research Approach

The project was divided into three primary tasks:

Task 1: To quantify the magnitude of savings, we estimate the baseline costs for procurement of 60 aircraft and associated engines, including spare engines, in FYs 2007–2009 under annual (single-year [SY]) contracting rules at the rate of 20 aircraft per year and compare the estimated SY prices to the negotiated multiyear (MY) contract price.

Task 2: To provide perspective on realism of savings, we assess the reasonableness of MYP cost savings by substantiating the sources of proposed savings and tracing them into the final negotiated contract.

We then compare these savings to the MYP prices minus the SYP estimates.

Task 3: To provide a historical context, we collect and report on the historical cost savings resulting from previous and ongoing aviation-related (aircraft and aircraft engines) multiyear procurement contracts authorized under provisions of 10 USC 2306b dating back to FY 1982.

How We Went About This Research

To accomplish these tasks, we reviewed existing literature on multiyear contracts of other aircraft programs and analyzed historical data. We also reviewed both the multiyear Lot 7, 8, and 9 proposals, and the Lot 7 single-year proposals submitted by the F-22A air vehicle and engine prime contractors. We visited and collected considerable information and historical data from the F-22A System Program Office and prime contractors—LM, Boeing, and Pratt & Whitney (P&W)—and major subcontractors involved in the production of the F-22A. Moreover, we visited and gathered information from the C-17, C-130, F/A-18E/F program offices, and the Institute for Defense Analyses (IDA). Figure 1.1

Figure 1.1
F-22A System Program Office and Contractor Locations Visited

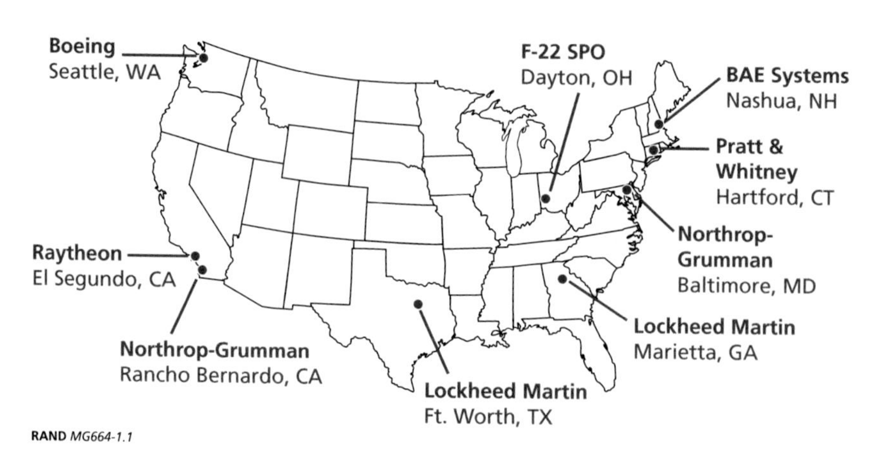

shows the F-22A vendors we visited and where they are located. Table 1.1 provides information about what is produced at each site.

Table 1.1
Contractor Locations and Roles in F-22A Production

Contractor and Location	Area of Expertise	Activities
Lockheed Martin, Marietta, Ga.	Forward fuselage assembly and final assembly and check-out	Lockheed's Marietta facility is responsible for a significant portion of the work on the F-22A. It oversees the weapon system integration; development and production of the forward fuselage, vertical fins, stabilators, wing and empennage leading edges, flaps, flaperons, and landing gear; and final assembly and flight testing. It also spearheaded "avionics architecture development and functional design, as well as displays, controls, the air data system and apertures."[a] In addition to the F-22A, Marietta produces the C-130J and is involved in a number of other programs, such as the modernization of the C-5 Galaxy.
Lockheed Martin, Ft. Worth, Tex.	Mid-fuselage fabrication and assembly	Lockheed's Ft. Worth facility is home to the headquarters of its Aeronautics Division, as well as the primary plant for the F-16 and JSF. Ft. Worth also has responsibility for "developing and constructing the mid-fuselage and armament; providing the tailored INEWS (Integrated Navigation and Electronic Warfare System), CNI (Communication, Navigation, and Identification), stores management systems and inertial navigation systems; [and] developing the support system."[a]
Boeing, Seattle, Wash.	Wing and aft fuselage fabrication and assembly	Boeing produces and assembles the wings and aft fuselage in Seattle under its Integrated Defense Solutions (IDS) division. In addition to these two parts, Boeing is responsible for "avionics integration and test; 70 percent of mission software; the pilot and maintenance training systems; and the life-support and fire-protection systems."[b] Boeing has a number of production centers near Seattle, mainly for its commercial aircraft. IDS is responsible for several other major military aircraft, including the F/A-18, the B1-B, and the F-15, at other locations.

Table 1.1—continued

BAE Systems, Nashua, N.H.	Electronic warfare (EW)	BAE Systems is the primary contractor for the F-22A Electronic Warfare suite. Nashua houses several facilities, including the headquarters for the BAE Electronics and Integrated Solutions Group (E&IS), and is an important site for the Electronic Warfare and Sensors divisions of E&IS. E&IS has a diverse range of civilian and military products, including the Joint Tactical Radio system and IFF Systems. The EW suite represents a small part of a business that includes supplying a wide array of EW systems for current aircraft as well as the suite for the F-35.
Raytheon, El Segundo, Calif.	Common Integrated Processor (CIP)	Raytheon produces the Common Integrated Processor in El Segundo, Calif., within its Space and Airborne Systems division. This division also builds the processors for the F-16 and F/A-18 and is contracted to provide the F-35's processor. These processors represent a small part of the product line, which includes a range of military and space sensors and a number of military GPS systems.
Northrop-Grumman (NG), Rancho Bernardo, Calif.	Communications, Navigation, Identification (CNI)	Northrop-Grumman is developing the CNI system for the F-22A under its Radio Systems business within the Network Communications Division. This division is generally focused on space systems, while the Radio Systems business focuses more closely on software for radios. The CNI appears to be a minor contract for the Network Communications Division.
Northrop-Grumman, Baltimore, Md., in a joint (JV) venture with Raytheon, Dallas, Tex.	AN/AGP-77 radar	The AN/AGP-77 is developed and produced as a joint venture between Northrop-Grumman's Electronic Systems sector (within its Aerospace Systems division) and Raytheon's Space and Airborne Systems sector. NG's division produces a wide array of military aircraft systems, including B-1, F-16 and AWACS radars, and is contracted to build the radar for the F-35. It also produces sensors for the Space Radar program and the SBIRS satellites. Raytheon produces a number of fire control and Airborne Electronically Scanned Array radar systems, as well as electronic warfare systems and mission computers for a number of weapon systems.

Table 1.1—continued

TIMET Corporation, Dallas, Tex.	Titanium	The TIMET Corporation is one of three major titanium producers in the United States. Its Dallas location is the world headquarters for its operations. TIMET supplies titanium for a wide variety of applications, from commercial aerospace to medical applications.
Pratt & Whitney, Hartford, Conn.	F119 engine (GFE)	The F119 engine is built just outside Hartford, Conn., by the Military Engines division of Pratt & Whitney, which is owned by United Technologies Corporation. This division also produces the engines for the F-15, F-16, and C-17 and will produce the F135 engine for the F-35 program. The commercial engines program is also based in the area.

[a] See http://www.lockheedmartin.com/wms/findPage do?dsp=fec&ci=15116&rsbci=15047&fti=0&ti=0&sc=400

[b] See http://www.boeing.com/defense-space/military/f22/index.html

Organization of Report

Chapter Two describes the statutory requirements and criteria for multiyear contracts and the benefits that typically result from such contracts. Chapter Three compares results of the RAND single-year procurement model with the negotiated multiyear contract prices and summarizes our findings. Chapter Four describes our categorization and substantiation process of the contractors' proposed initiatives to achieve savings due to multiyear contracting. Chapter Five reviews other aircraft programs that have used multiyear contracts, and Chapter Six provides our review and assessment of savings initiatives proposed by the prime contractors and some of the major vendors.

The report also includes five appendixes. Appendix A provides detailed analysis of the F/A-18E/F cost data in an attempt to validate the claimed savings; Appendix B summarizes reports by the Government Accountability Office (GAO), IDA, and RAND on issues related to multiyear procurement; Appendix C provides an empirical analysis of the tail-up costs, and Appendix D is a summary of the 2006

IDA business case analysis (BCA) on the F-22A multiyear savings. Appendix E provides some detailed information about multiyear, fixed-wing aircraft procurement contracts let since 1995.

CHAPTER TWO

The Basics of Multiyear Contracts

This chapter discusses multiyear contracts in general, describing their requirements and the benefits they offer. It also addresses some aspects of the F-22A multiyear contract. Those familiar with multiyear contracts and the F-22A contract may wish to skip this chapter.

Multiyear Contract Requirements

Multiyear contracts are permitted by 10 USC 2306b, which defines a multiyear contract as follows:

> For the purposes of this section, a multiyear contract is a contract for the purchase of property for more than one, but not more than five, program years. Such a contract may provide that performance under the contract during the second and subsequent years of the contract is contingent upon the appropriation of funds and (if it does so provide) may provide for a cancellation payment to be made to the contractor if such appropriations are not made.[1]

In general, Congress must authorize the use of a multiyear contract whenever one or more of the following events occur:

[1] 10 USC 2306b, subparagraph (k).

- The contract will exceed $500 million for supplies or $572.5 million for services.
- It will employ economic order quantity procurement in excess of $20 million in any one year.
- It will employ an unfunded contingent liability in excess of $20 million.
- It will involve a contract for advance procurement leading to a multiyear contract that employs economic order quantity procurement in excess of $20 million in any one year.
- It will include a cancellation ceiling in excess of $100 million.[2]

Criteria for Using Multiyear Production

Title 10, USC 2306b, stipulates that the head of an agency may enter into multiyear contracts for the purchase of property to the extent that funds are available for obligation when each of the following criteria can be met:

1. That the use of such a contract will result in substantial savings of the total anticipated costs of carrying out the program through annual contracts.
2. That the minimum need for the property to be purchased is expected to remain substantially unchanged during the contemplated contract period in terms of production rate, procurement rate, and total quantities.
3. That there is a reasonable expectation that throughout the contemplated contract period the head of the agency will request funding for the contract at the level required to avoid contract cancellation.
4. That there is a stable design for the property to be acquired and that the technical risks associated with such property are not excessive.

[2] See FAR Part 17.104(c).

5. That the estimates of both the cost of the contract and the anticipated cost savings through the use of a multiyear contract are realistic.
6. In the case of a purchase by the Department of Defense, that the use of such a contract will promote the national security of the United States.

This report does not address all the criteria listed above, but it provides an assessment of the cost savings associated with the F-22A MYP contracts for the air vehicles and engines, in contrast to three single-year contracts for each.

Basic Components of the F-22A Contract

The USAF plans to award a firm-fixed-price (FFP) multiyear contract to the Lockheed Martin–Boeing team for three lots of 20 F-22A aircraft (60 total) and associated support and a separate FFP multiyear contract to P&W for three lots of 40 F119 engines plus spare engines (a total of 133 engines) and associated support in July 2007.[3] The contracts will cover funding authorized and appropriated in FYs 2007–2009. Lot 7 of the program will be primarily funded with FY 2007 funds, except as discussed in the Funding for Production section below. The overall funds allocated to the multiyear contracts combined will be just over $10 billion. These two contracts represent about 70 percent of the total funding appropriated (FY 2007), budgeted (FY 2008), and programmed (FY 2009) for the F-22A program.

Figure 2.1 shows the allocation of the total funding of approximately $15 billion during FYs 2007–2009. As the figure shows, 70 percent of the budget is allocated to the multiyear contract, 19 percent to the F-22A modernization program, 7 percent to operation and maintenance (O&M), 2 percent to military construction (MILCON),

[3] About 1 percent of the multiyear contracts will be cost-plus contract line items.

Figure 2.1
F-22A Funding, FYs 2007–2009

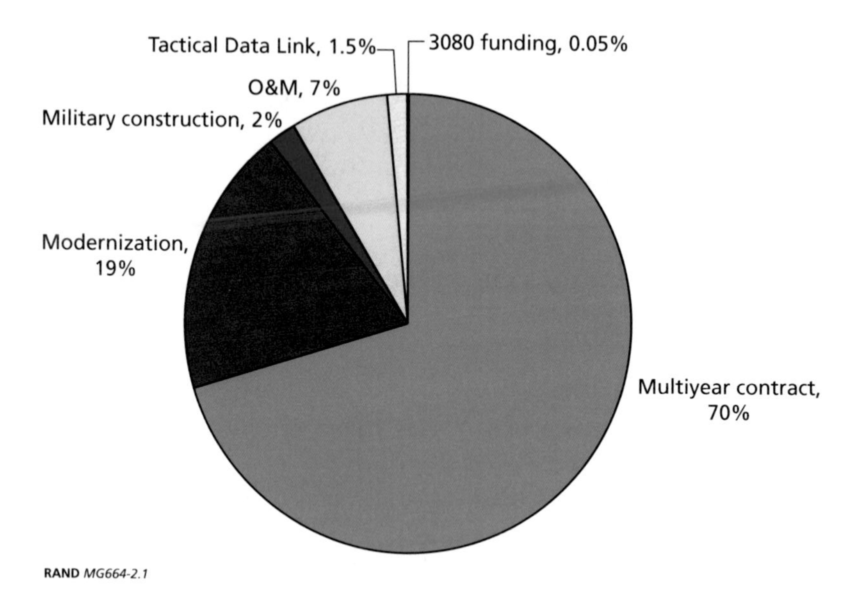

1.5 percent to the Tactical Data Link system,[4] and 0.05 percent to 3080 funding[5] to cover chaff and flares procurement.

Content Included in the F-22A Multiyear Contracts

The multiyear contracts will include all production activities and materials for Lots 7–9, including the delivery of the 60 aircraft and 120 installed engines plus 13 spares, aircraft maintenance equipment and information systems, production support, training equipment and installation at operational USAF bases, useful loads, management of diminished manufacturing sources, tail-up costs,[6] and some initial spares. It also includes two categories called Program Support Sustain-

---

[4] The Tactical Data Link is the communications system that provides digital information to the pilot about enemy and friendly aircraft, and such. Information can be transmitted and received across various units and platforms, including NATO, the Navy, and the Army.

[5] 3080 funding is classified as "Other Procurement" under the U.S. Air Force Budget.

[6] *Tail-up costs* are those additional costs associated with the shutting down of a production line. These are described in more detail in Appendix C.

ment (PSS) and Program Support Producibility (PSP), which involve support for processing aircraft changes, configuration management, proposal preparation, and other sustaining activities not directly associated with hardware delivery.

F-22A Program Content Not Included in the Multiyear Contract

Two major areas outside of the multiyear contract purview that will be on separate contracts are the following:

- The Program Agile Logistics Support (PALS) contract, which covers the basic interim contractor-furnished logistics support for the initial deliveries of aircraft and hardware
- The Raptor Enhancement Development and Integration (REDI) contract, which includes the development and test efforts for additional or enhanced capabilities for the F-22A and is funded with research, development, test, and evaluation (Appropriation 3600) funds only.

Also outside the multiyear contract are activities to shut down production and close out the program after the last delivery (to be covered under separate contracts with the prime contractors), sustainment activities (funded with O&M (Appropriation 3400) dollars), other government costs, and construction of new buildings at operational bases (funded with military construction (Appropriation 3300) dollars).

Funding for Production

F-22A production is funded in three categories, all under the Aircraft Procurement, Air Force (Appropriation 3010) funds. These are the basic procurement funds, advance procurement funds, and economic order quantity funds. The reason for these divisions stems from a longstanding policy called "full funding," wherein Congress will only authorize and appropriate procurement funds to buy useful, complete end items, not parts of units. DoD Regulation 7000.14-R explains the overall process as follows:

> Procurement of end items shall be fully funded, i.e., the cost of the end items to be bought in any fiscal year shall be completely included in that year's budget request. However, there are occasions when it is appropriate that some components, parts, material, or effort be procured in advance of the end item buy, as authorized, to preclude serious and costly fluctuation in program continuity or when items have significantly longer lead times than other components, parts, and material of the same end item. In these instances, the long lead-time material or effort may be procured with advance procurement funds, but only in sufficient quantity to support the next fiscal year quantity end item buy (except for economic order quantity procurement of material to support a multiyear procurement), and only to buy those long-lead items necessary to maintain critical skills and proficiencies that would otherwise have to be reconstituted at significantly greater net cost to the Government. When advance procurement is part of a program, the cost of components, material, parts, and effort budgeted for advance procurement shall be relatively low compared to the remaining portion of the cost of the end item. Because such use of advance procurement limits the MDA's [Milestone Decision Authority's] flexibility, this acquisition technique shall be used only when the cost benefits are significant and only with approval of the MDA.[7]

The Federal Acquisition Regulation (FAR), part 217, also addresses the two exceptions:

> "Advance procurement" means an exception to the full funding policy that allows acquisition of long lead time items (advance long lead acquisition) or economic order quantities (EOQ) of items (advance EOQ acquisition) in a fiscal year in advance of that in which the related end item is to be acquired. Advance procurements may include materials, parts, components, and effort that must be funded in advance to maintain a planned production schedule.[8]

---

[7] DoD Instruction 5000.2-R, April 5, 2002, Chapter 2.

[8] FAR, Part 217.103, Definitions.

> Funds appropriated for any fiscal year for advance procurement are obligated only for the procurement of those long-lead items that are necessary in order to meet a planned delivery schedule for complete major end items that are programmed under the contract to be acquired with funds appropriated for a subsequent fiscal year (including an economic order quantity of such long-lead items when authorized by law (10 USC 2306b(i)(4)(b)).[9]

Economic order quantity funds, however, are unique to multiyear contracts. The philosophy behind EOQ funds is that ordering certain materials or parts for the entire production run envisioned in the multiyear contract can produce savings by eliminating annual buys and production line setups and terminations. Under EOQ, savings are the key feature, rather than schedule, and funding is normally used for recurring costs. Thus, buying new production tooling with EOQ funds would normally not be permitted. For the F-22A program, $300 million in EOQ funds were appropriated in FY 2007, with no further funds programmed for FYs 2008 and 2009. The FAR also lists several restrictions on the use of EOQ funds, based on direction from 10 USC 2306b.

Cost Reduction Initiatives (CRIs)

Another unique feature of the F-22A multiyear contract is the lack of funded cost reduction initiatives. Some multiyear programs set aside funds to support engineering efforts to improve production processes, improve or replace tooling, or change designs to facilitate production. Such efforts are typically one-time events and tend to occur at either the prime contractor or major subcontractor level, rather than at the supplier level. CRIs are generally nonrecurring in nature, so EOQ funds are normally not used to pay for them. Because only three more lots of F-22A production are currently planned, virtually no CRI efforts are specifically funded in Lots 7 through 9.

---

[9] FAR, Part 217.172(e)(6).

Program Acceleration

With the emphasis on savings resulting from the multiyear contracting approach, one might ask how further savings could be achieved besides using EOQ funds and possibly pursuing additional CRIs beyond those implemented earlier in the F-22A program. Certainly, accelerating the entire 60-aircraft production was an option at one point. Tooling is in place at the prime contractors as well as major vendors that would allow as many as 30 F-22As to be produced per lot, so the 60-aircraft program could be completed in two rather than three years. This certainly would produce savings by spreading the fixed overhead over greater quantities in two years and then shutting the production facilities down, assuming no further aircraft are anticipated. The major drawback to this is the requirement for moving about $1.5 billion into FYs 2007 and 2008 from FY 2009. This is unlikely, given budget constraints and the need for major supplemental funding in FY 2007. This acceleration is probably impossible now (mid-2007) because of schedules and other commitments already set in place for producing only 20 aircraft per year.

Multiyear Contract Modifications Possible

Our evaluation of the multiyear contracts was made on the stated terms of the "instant contracts," i.e., the contracts to be awarded in July or August of 2007. The contracts are firm-fixed-price contracts, which means that any decreases or increases in production costs are basically born by the three major contractors.[10] There will be an economic price adjustment (EPA) clause in the multiyear contracts. However, the government can legally modify any contracts after award. Three areas that will likely result in contract modifications for the F-22A multiyear contracts are the following:

- The addition of training systems not specified in the instant contract
- Solutions to the corrosion problems found in certain areas of the airframe

---

[10] See footnote 3 on the cost-plus portion of the multiyear contracts.

- Any engineering or design change proposals resulting from the modernization program (REDI contract) efforts that, for safety or operational capability reasons, become desirable to implement on some or all of the 60 aircraft and engines produced under the multiyear contracts.[11]

Stability in F-22A Configuration Is Paramount During the MYP Contract

The foundation on which all multiyear savings initiatives are based is a stable configuration of the F-22A during production Lots 7–9. The multiyear contract is based on the Lot 6 configuration continuing throughout the remaining 60 aircraft with no substantial changes. If the configuration must be changed, the savings forecast could be decreased during Lots 7–9. The effect on forecast savings would be a function of the nature of the modifications. In certain cases, such as the insertion of straightforward, "plug-and-play" substitutions of parts (i.e., the form, fit, and function remain the same), modifications might not result in significant cost increases if the design engineering and tooling are insignificant and the parts have not already been ordered under an EOQ arrangement.

More significant configuration changes could require reengineering of parts, changes in tooling, reordering of replacement parts already purchased as EOQ, renegotiating contracts for changed work, and so forth. Thus, without a solid commitment to retaining the Lot 6 configuration throughout the next three multiyear lots, the forecast savings could be jeopardized and the magnitude of the savings would depend on the nature of the configuration changes. The savings forecast for stability in, or elimination of, diminished manufacturing sources (DMS) management is a good example of how savings from configuration stability and ordering parts in advance can produce savings under a multiyear contract. A stable configuration can reduce the

---

[11] The F-22 SPO has highlighted in a number of occasions that the technologies developed through the modernization program will be retrofitted to the aircraft procured in Lots 7, 8, and 9.

need for recurring engineering costs for design efforts, tooling changes, manufacturing planning, and the number of configuration control board–like activities.

CHAPTER THREE

Estimating Single-Year Procurement Price

Introduction

One of the more challenging tasks in assessing the magnitude of savings due to a multiyear procurement strategy is evaluating the costs of the alternative single-year strategy (the path not taken). That is, if a multiyear contract is awarded, then the actual costs under an equivalent series of single-year contracts will never be known with certainty. Similarly, if procurement proceeds with single-year contracts, then the actual savings that might have occurred under a multiyear contract remain only theoretical (or as proposed values). In part, this ambiguity increases further because of changes in the actual execution of the program under either scenario (shifting funding earlier, investing in EOQ, cost reduction initiatives, etc.), such that a direct, after-the-fact extrapolation between multiyear and single-year scenarios becomes difficult—even with existing, actual cost data.

The costs of the path not taken can, of course, be *estimated*. But such estimates are subject to the same uncertainties as any estimate—they are not exact forecasts. In this chapter, we estimate the equivalent price[1] of the F-22A procurement for FY 2007 through FY 2009 as if it had been three independent single-year procurements for 20 aircraft

[1] Here it is important to distinguish between price and cost as generally used in defense acquisition circles. *Cost* formally corresponds to the amount paid for an item exclusive of profit or fee. *Price* is the full amount paid to the contractors, including all fees and profit. In

each, along with associated engines and spares. These single-year prices will serve as the baseline from which we evaluate the multiyear savings for FYs 2007–2009.

Scope of the Estimate

In our estimating approach, we divide the procurement price for the F-22A program into six cost elements. The definitions of the elements in Table 3.1 are taken from the definitions provided by the program office.[2]

The multiyear proposals do not contain the full procurement price, thus complicating the multiyear savings analysis somewhat. For example, the LM and Boeing multiyear proposals cover all of TPC and a portion of PSO and PSAS program values. However, none of the OGC or OPC prices are included. The costs of any emerging requirements or the modernization efforts are also not covered in the multiyear proposals.[3] The multiyear proposal for P&W covers the price for all engines to be installed for Lots 7 through 9 but only the spare engines for Lots 8 and 9.[4] To keep the multiyear/single-year comparison on an equivalent basis, our single-year estimate will have the same program scope as the multiyear proposals. Therefore, the reader must be careful not to confuse the values presented in this chapter with those of other documents or the amount planned in the Air Force budget. The values shown are for the majority of the total F-22A procurement price for Lots 7, 8, and 9, but not the entire amount.

---

this chapter, we estimate the *price* of the single-year procurement. While we use the terms interchangeably in the chapter, all values shown are formally prices.

2 F-22 Program Office, 2006.

3 See Chapter One for a further explanation of multiyear contract content.

4 Lot 7 engine spares are expected to be part of the field support and training (FS&T) contract. However, we include the spares units as part of the overall production cost improvement curve.

Table 3.1
Program Office Definitions of F-22A Production Cost Elements

Cost Element	Description
Target Price Curve (TPC)	All costs associated with production of aircraft, excluding sustaining labor that cannot be uniquely identified with a particular aircraft or lot (flight hardware).
Product Support Other (PSO)	Includes the costs for trainers, peculiar support equipment (PSE), Integrated Maintenance Information System (IMIS), Mission Support System (MSS), Operational Debrief System (ODS), alternative mission equipment (AME), useful loads, rate tooling, and diminishing manufacturing sources (DMS).
Program Support Annual Sustaining (PSAS)	Costs for sustaining engineering and program management support labor not included in TPC costs and not directly related to the aircraft build.
Performance-Based Agile Logistic Systems (PALS)	Costs for contractor logistic support for the air vehicle and engine, primarily production initial consumable spares (ICS) and initial spares. Some of the initial spares are included in the multiyear contract; we identify these as initial spares in the rest of the chapter.
Propulsion	All costs associated with the production of engine installs, whole engine spares, engine spare parts & modules, and engine field support & training (FS&T). The propulsion-related costs are broken out separately because the engine contract is separate from the contract for the rest of the aircraft.
Other government cost (OGC)/other production cost (OPC)	Government furnished property (GFP), common support equipment (CSE), expendables, mission support, risk, etc.

Estimating Approach

Data Sources

All estimates use some form of data, typically historical, to forecast future values. Fortunately, the F-22A program has a defined procurement history on which to base an estimate for Lots 7–9. Therefore, we can use those data directly for the most part and not have to adapt or adjust data from other fighter programs. The program has been in production since 1999, beginning with the first two lots of production representative test vehicles (PRTVs). As of Lot 6 (the FY 2006 procurement), 123 total aircraft and associated engines and spares are on contract. Production is currently under way for Lot 5 and 6 aircraft and engines, and long-lead activities for Lot 7 are just beginning.

We use this program experience to forecast the TPC and propulsion prices for Lots 7 through 9. The data are based on actual certified costs for Lots 1 through 4, partial returned costs for Lot 5 (including some estimates at completion), and the negotiated values for Lot 6. We have excluded the earlier development EMD and PRTV lot costs because they are not representative of the way the aircraft is currently being produced or its current configuration. Labor rates and factors are based on either forward priced rate agreements (FPRAs) or projected values provided by each firm. Material and equipment escalation is based on Bureau of Labor Statistics (BLS) data (described later). We assumed the same profit level as in the negotiated settlement.

For the PSAS, PSO, and other non-TPC prices, we use another data source because of the scope of the estimate issues discussed above. We estimate the single-year price for these elements by taking the multiyear negotiated settlement values and adding the identified savings (shown in Chapter Four) to the settlement for Lots 8 and 9. For Lot 7, we used the negotiated single-year prices directly.[5]

Throughout this chapter, we present costs and prices in then-year dollars (TY$) and not constant or budget-year dollars. That is, the costs and prices are those anticipated to be paid at the time the payment is made—including any price escalation that occurs. The main reason for the use of TY dollars is that we can make direct comparisons with the contractor's proposals, negotiated values, and budget documents, which are in TY dollars only. This comparison is the main objective of our study. By using only TY dollars, we avoid distorting these numbers. Furthermore, it avoids having different values for the same negotiated prices (i.e., then-year and fixed-year values) that could lead to confusion if misquoted.

---

[5] As part of the MYP contract strategy, the F-22A SPO asked the contractors to present proposals for both a single-year Lot 7 award and a three-year multiyear award. Both proposals were fully negotiated between the SPO and the contractors so Lot 7 could proceed should a multiyear award not be approved.

In addition, placing the contractor values on a fixed-dollar basis would require us to (a) time-phase the expenditures by year and (b) select an appropriate deflator. Performing (a) correctly is not obvious, because we know the pattern of spending changes subtly as a result of its multiyear nature (some spending gets shifted to an earlier period). We do not have enough detail on spending patterns to discern differences between multiyear and single-year plans. Issue (b) is more problematic in terms of a choice in the deflator: Which one should we pick? There are several possible deflators. We are reluctant to use the DoD deflators because they do not incorporate some of the very recent price inflation (such as that observed with specialty metals). Thus, our savings value would be distorted. We could use a BLS-based index (as we did for our projection of the single-year equipment and material prices), but again our numbers would differ from the "actual" values assumed and could thus introduce distortions. The labor, equipment, and material inflation in the contractors' negotiated values are based upon FPRAs and vendor quotes, not on any particular escalation index.

Level of Detail

A commonly employed estimating method for programs in production is an approach in which costs are generated at a very detailed activity or part level and then summed to a total amount. For typical estimates, hundreds to thousands of individual items may be addressed. Given the constraints for this study, we did not have the time to build such a highly detailed price model. Rather, we aggregated the historical data at a higher level of indenture to populate our model. A higher-level approach is sufficient for this analysis because we do not need to track costs at a detailed level (i.e., we are not generating a control estimate) and we are not making changes in assumptions that would disproportionately affect particular items (at least for the single-year contracts). For example, we are not changing the aircraft configuration from that committed for Lot 6. So, there are no "baseline" changes to specific cost elements.

The data were aggregated as follows: For TPC costs, the data were aggregated by firm (i.e., Lockheed Martin and Boeing) as either labor or procurement (material) items. Labor was further segregated into touch (manufacturing) and engineering by manufacturing site (i.e., Fort Worth, Marietta, Palmdale, and Seattle). Non-TPC prices were estimated at the total price level by contract line item number (CLIN). The propulsion costs were based on unit price data for the whole engines and have not been broken out by labor hours and material due to the limited data available. We did not include any costs for OGC or OPC because they are not part of the multiyear contract. Table 3.2 summarizes the cost detail by cost element.

Table 3.2
Level of Cost Detail in RAND SYP Price Model

Cost Element	Total Price	Labor	Material
TPC		Engineering and touch by site	By firm
PSO	X		
PSAS	X		
Initial spares	X		
Propulsion	X		
OGC/OPC	Not included	Not included	Not included

Forecasting Method

As described above, we estimate the non-TPC single-year prices directly from the values negotiated between the USAF and contractors. We also use the negotiated values for the unique item identification marking (UID) costs, which are an additional requirement for Lots 7 to 9. For the remainder of the TPC subelements, we use a series of cost improvement curves (CICs) with a rate term to forecast the prices for Lots 7–9. The CICs take the following form:

$$C_n = C_1 \times n^{\ln(b)/\ln(2)} \times r^{\ln(c)/\ln(2)} \qquad 3.1$$

where

C_n is the average cost (or hours) for the lot
C_1 is the cost (or hours) for the first unit
n is the lot unit number midpoint
b is the unit cost improvement slope[6]
r is the number of units procured in the lot
c is the rate slope.

Normally, one would determine the rate and improvement slopes simultaneously through multivariate analysis by regressing the log of the historical values versus the log of the lot unit midpoint and log of the lot size. However, for the actual F-22A production history, the rate and unit midpoint values are so highly correlated that such an approach is invalid. Instead, we applied Equation 3.1 to the total procurement cost for several other historical aircraft programs that did not have the same correlation problems.[7] The average value for the rate slope was 89 percent. We fixed the rate slope for the F-22A CICs to 89 percent in Equation 3.1 and used regression to determine the cost improvement slopes. For the propulsion category, we fit Equation 3.1 to historical data for F100 engine (-100/-200 and -229) and used the resulting rate slope. The average rate slope for these engines was approximately 97 percent.

The most important part of the single-year cost analysis is to select the portion of the production Lot 1–6 cost improvement curve over which to determine the likely improvement slope for Lots 7–9. To bracket our forecast of the single-year price, we used three assumptions:

---

[6] See Fisher, 1970.

[7] These aircraft programs were A-10A/B, A-6E, A-7E, AF-2W, AV-8A, AV-8B, C-130H, C-2AR, C-9, EA-6B, EC-130, F-14A, F-15E, F-16A/B, F-16C/D, F-4B, F-4J, KC-130J, P-2H, SMALL VCX (C-37), T-33B, T-34C, T-45A, and TA-4J.

A. *The trends in cost experienced during Lots 1–6 will continue through Lots 7–9.*[8] This is the most aggressive assumption (i.e., it should yield the lowest lot prices) because it includes all cost reduction initiatives embedded in the data. Basically, this case assumes the continuation of the typical cost improvement ("learning") curve for the last three lots. Both the contractors and the SPO feel that this assumption overstates the cost improvement that might realistically occur for Lots 7–9 because no additional funding is planned to implement further cost reduction initiatives for those lots. However, the normal process improvements are anticipated to continue for those lots. Compensating for these reductions are the configuration changes and performance improvements that were also included as part of the Lot 1–6 price changes.
B. *The trends for Lots 5 and 6 will continue through Lots 7–9.* In this assumption, a new CIC was developed using the same cumulative quantities and lot midpoints as above, but only the Lot 5 and 6 costs were used in the regression. The new CIC was used to predict costs for Lots 7–9. These resulting costs were adjusted for the change in annual production rate from Lot 6 to Lot 7 and beyond. This assumption is probably the most realistic one. The data for Lots 5 and 6 already incorporate much of the cost reduction initiatives in their baseline, so one presumably will not be forecasting as much improvement in future lots compared with the Lot 1 to 6 assumptions. Furthermore, Lots 5 and 6 (particularly Lot 6) share a configuration that is most similar to that planned for Lots 7–9. Also, Lots 5 and 6 reflect some of the unusual recent price escalation for materials that all the firms have experienced—particularly for metals. Last, the CIC slope (or rate of improvement) reflects recent improvement

[8] The F-22A program had invested about $630 million in production cost reduction initiatives in Lots 1 through 6. Ninety-five percent of these funds were spent in Lots 1 through 4 and the remaining 5 percent in Lots 5 and 6.

trends and not those based on the initial lots where more productivity improvement program (PIP) funding was available and greater opportunities for improvement existed. The improvement slopes for this assumption were generally flatter compared with the first assumption.

C. *Lot 6 cost data are the most indicative of likely costs for the next three lots.* This assumption extrapolates the future prices from Lot 6 values and assumes no further cost improvements. Lot 7–9 costs were adjusted for the lower annual production rate and for inflation. This is the most conservative assumption. It produces the highest single-year estimates and serves as the upper bound for our estimate of single-year prices.

Other Adjustments

One specific adjustment that must be made to our forecast of the SYP price for the F-22A is a "tail-up" allowance. *Tail-up costs* are the additional costs associated with the inefficiencies normally experienced as the production ends. They do not include the cost of formally closing the line (such as preserving tooling and remediation costs). Since the basic activities are the same under either a multiyear or annual contracting strategy, we assume these costs will be the same in either case. We estimate the tail-up cost using historical data on aircraft procurement. We have assumed a price shift upward of 7.4 percent for Lot 9 recurring labor and materials. The analysis details of this adjustment are provided in Appendix C.

Recently, there have been dramatic increases in the spot market prices of specialty metals, such as nickel and titanium. Figure 3.1 shows the relative price of titanium mill shape products based on the Producer Price Index from the Bureau of Labor Statistics—BLS (WPU102505). The relative prices are plotted for each calendar year (the corresponding lot that is affected is shown in parentheses). Note the two-year offset between the calendar year and FY lot, owing to the long-lead nature of titanium for the F-22A production. From 1999 to 2004, the relative prices increased about 18 percent. However, the increase in price between 2004 and 2006 is about 135 percent (the price more than doubled in three years).

Figure 3.1
Relative Annual Price of Titanium Mill Products

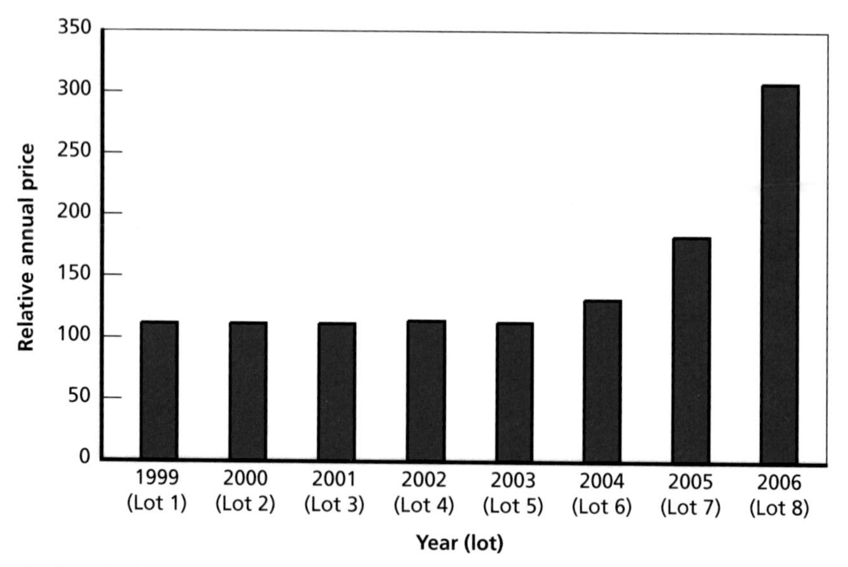

SOURCE: BLS.
NOTE: 1982 = 100.
RAND *MG664-3.1*

To account for this unusual escalation in titanium cost, we added an additional escalation for that portion of the material price. Lockheed Martin Aerospace and Boeing supplied the weight of titanium that was not covered by their long-term agreements with suppliers (and hence was affected by this change in the spot market price). The two firms also provided the dollars per pound (spot price) for titanium they have paid or expect to pay for Lots 5–9. The product of the two values gives the *estimated* cost of titanium by lot. Based on the Lot 5 titanium cost, we used the OSD deflator (our general material escalation index) to project a *theoretical* titanium lot cost under normal conditions. The difference between the *estimated* and the *theoretical* cost is our titanium escalation adjustment. To avoid escalating twice, we removed the escalation adjustment from Lot 6 material prices.

Price Model Structure

The price model structure is shown in Figure 3.2, as implemented in the program called Analytica. Based on the prior production quantities and the future lot quantities (rectangles on left), the model applies Equation 3.1 to each of the subelements for TPC (touch labor, engineering labor, material, and total price). The calculation separates the labor and material price calculations into separate calculations (in dark blue)—although both model paths work similarly. From these calculated labor and material prices, TPC and propulsion lot prices are calculated. The tail-up factor is added to the TPC costs. These values are next added to the "Non-TPC Price" (e.g., PSAS, PSO, etc.) to determine a contract price per lot and an average unit price (AUP). The parallelograms at the bottom show dimensions over which the data are

Figure 3.2
High-Level Single-Year Price Model Structure

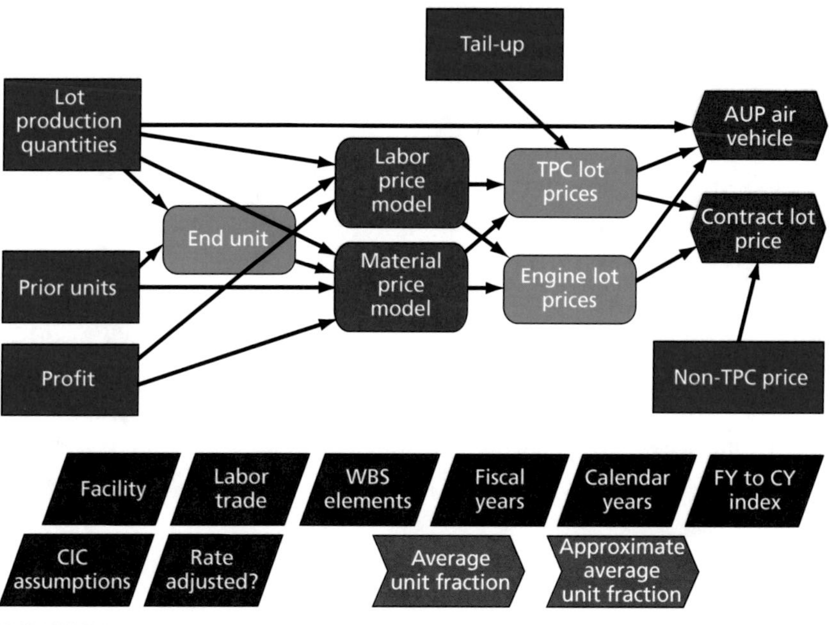

RAND *MG664-3.2*

organized (e.g., the work breakdown structure, fiscal years, calendar years, labor type).

Figure 3.3 shows the detail for the labor price model. Based on the production quantities, Labor C_1s, CIC slopes, and rate factors, the model calculates the total labor hours for each lot by applying Equation 3.1. Also added to the labor hours are factored hours (e.g., QA (quality assurance) and support hours) that are based on the direct touch or engineering labor hours. There is also a provision for shifting the baseline for change such as configuration shifts after Lot 6; however, this feature is unused in the current analysis because no such shifts are planned. Using the hours by lot, the model next calculates the hours by calendar year using a work spread in time based on data from the firms. These hours by calendar year are next priced by multiplying the appropriate wrap rate (fully burdened labor cost including fee) to determine a labor price by calendar year. The last step is to convert the calendar year prices back to fiscal year prices.

The material price submodel, shown in Figure 3.4, is structured similarly to the labor submodel. Again, based on C_1, slopes, and rate factors, the model generates constant-dollar cost by lot (FY 2000$). The lot costs are then spread over the appropriate fiscal years using a spending profile and escalated to TY$. Finally, the TY$ costs are converted back to FY lot prices by applying a factor and fee. The special titanium escalation adjustment is added at this point as well. The main difference between the labor and material submodels is that C_1s for material costs are in constant FY 2000$ and are later escalated to TY$ using a procurement deflator based on data from the BLS. For the TPC cost element, we used the Aircraft Manufacturing from the Industry series (PCU336411336411). For the propulsion cost element, we used the Aircraft Engine and Parts series (PCU336412336412). We extrapolated these series forward (in time) using an exponential time-series regression fit to the data between 2000 and 2006.

Figure 3.3
Labor Price Submodel Diagram

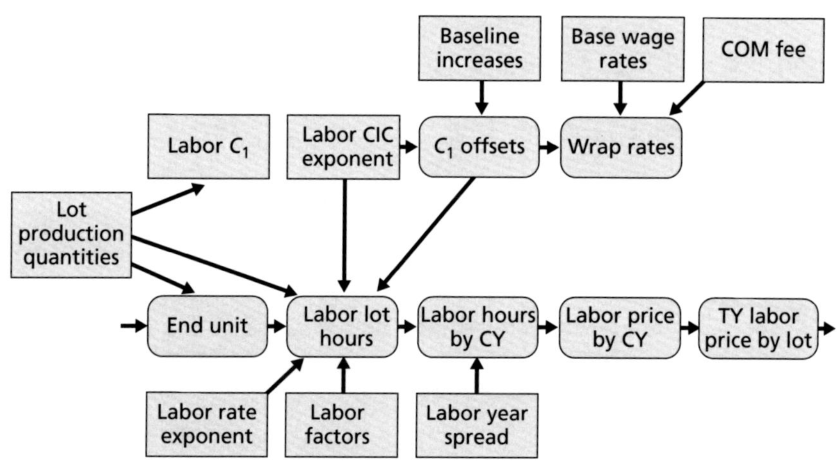

RAND *MG664-3.3*

Figure 3.4
Material Price Submodel Diagram

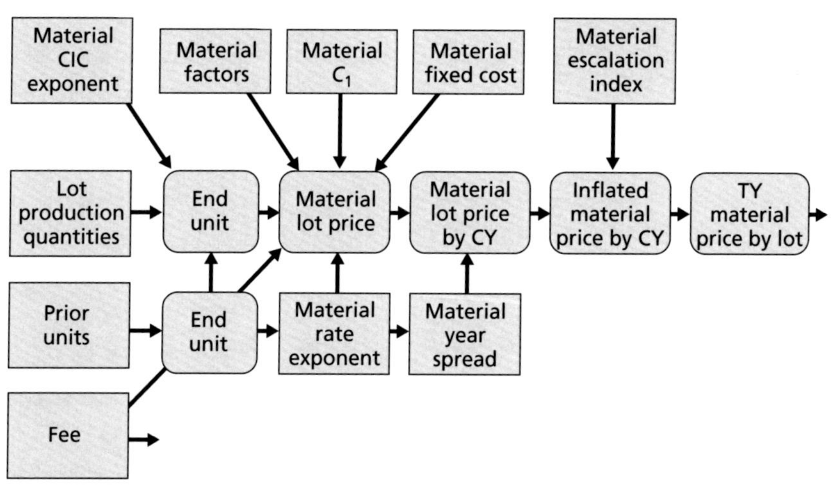

RAND *MG664-3.4*

Results

Using the data sources and input values described above, the model generates the overall contract price for the single-year equivalent of the multiyear work scope by lot. Table 3.3 summarizes the total contract price using the three CIC assumptions for all three firms.

As anticipated, the Lot 1–6 assumption forms the low end of the range for the total cost, whereas the Lot 6–only assumption forms the upper end. The range is roughly $9.0 to $9.3 billion with a baseline of approximately $9.1 billion. In no way should the range be interpreted as an uncertainty estimate or be seen as implying any degree of statistical confidence. The range of results merely displays the sensitivity of the estimate to differing CIC assumptions. Also, the numerical precision of Table 3.3 overstates the accuracy of the estimate. We report four significant figures to place the overall savings in context (which is in the hundreds of millions) so that rounding issues are minimized.

In Table 3.4, we display the equivalent of an average unit hardware price. We define the hardware cost in this case as the sum of the TPC (including UID nonrecurring cost) and Propulsion prices (excluding spares). The average value ranges from $126 to $145 (TY $millions) and depends on the CIC assumptions and fiscal year.

Table 3.5 shows the contract prices by cost element for the CIC assumptions of Lots 5 and 6. Again, the reader is cautioned that the precision of the numbers in the table does not reflect accuracy or uncertainty in the numbers. We have rounded to three significant figures.

Table 3.3
Estimated Total Contract Price for Single-Year Equivalent to Multiyear Scope for Boeing, LM, and P&W (TY $billions)

CIC Assumptions	Lot 7	Lot 8	Lot 9	Total
Lots 1–6	2.868	2.979	3.105	8.952
Lots 5–6	2.873	3.024	3.192	9.089
Lot 6	2.889	3.100	3.331	9.320

Table 3.4
Estimated Average Unit Hardware (TPC + Propulsion) Price (TY $millions)

CIC Assumption	Lot 7	Lot 8	Lot 9
Lots 1–6	125.9	126.5	134.4
Lots 5–6	126.2	128.7	138.6
Lot 6	127.0	132.4	145.4

Table 3.5
Estimated Single-Year Contract Price by Cost Element for B: Lot 5 and 6 CIC Assumption (TY $millions)

Cost Element	Lot 7	Lot 8	Lot 9	Total
TPC	2,131	2,160	2,346	6,637
PSO	195	182	129	506
PSAS	147	156	166	469
Initial spares	0	62	41	103
Propulsion	400	465	510	1,375
Total	2,873	3,024	3,192	9,089

NOTE: Propulsion costs for Lot 7 do not include the five spare engines.

We compare our Lot 7 single-year estimate to the value for the negotiated single Lot 7 as a check on the reasonableness of our assumptions and inputs. Table 3.6 summarizes the comparison at the total level for Lot 7 only. The negotiated Lot 7 value is barely distinguishable from our SYP estimates.

Table 3.6
Lot 7 Comparison of RAND Estimated SYP Price to Negotiated Proposal (TY $millions)

CIC Assumption	Lot 7
Lots 1–6	2,868
Lots 5–6	2,873
Lot 6	2,889
Negotiated Lot 7	2,866

The point of the SYP estimates is to produce an estimate of the MYP savings by subtracting the MYP negotiated values for each year from the corresponding SYP estimates for each assumption. Table 3.7 shows the savings by lot for each of the three assumptions. The range of savings is approximately TY $274–643 million.

Table 3.7
Estimated Multiyear Savings by CIC Assumption and Lot Number (TY $millions)

CIC Assumption	Lot 7	Lot 8	Lot 9	Total
Lots 1–6	43	108	123	274
Lots 5–6	48	153	210	411
Lot 6	65	229	349	643

Summary

We have estimated the single-year equivalent annual prices for the multiyear proposal scope based on F-22A program history data, negotiated proposal values, and differing CIC assumptions. The single-year price for Lots 7–9 is somewhere between TY $9.0 billion and TY $9.3 billion, depending on CIC assumptions. Our estimate for Lot 7 is consistent with the single-year Lot 7 negotiated values. The difference between our single-year estimates and the negotiated multiyear contract suggests a multiyear savings of between $274 million and $643 million, with $411 million as our most realistic estimate of the savings.

CHAPTER FOUR

Categorization and Substantiation of F-22A Multiyear Contractor-Proposed Savings

This chapter discusses the key initiatives developed by Lockheed Martin, Boeing, and Pratt & Whitney that are expected to produce savings for the F-22A multiyear contract. It addresses

- factors that lend themselves to quantification of forecast savings
- issues that can enhance savings but are more qualitative than quantitative in nature
- areas that are unlikely to result in savings under a multiyear contract in contrast to a series of single-year production contracts for Lots 7 through 9.

Evaluation of the Proposed Savings Initiatives

As part of the pre-award activities for the Lot 7–9 multiyear contract, the USAF asked Boeing, Lockheed Martin, Pratt & Whitney, and their subcontractors and suppliers to develop cost savings initiatives that could be implemented in the multiyear contract. The RAND team visited Boeing (Seattle, Washington), LM (Marietta, Georgia), P&W (Hartford, Connecticut), and their four main subcontractors: Northrop-Grumman Network Communications Division (Rancho Bernardo, California), Raytheon Space and Airborne Systems (El Segundo, California), BAE Systems (Nashua, New Hampshire), and the joint venture team of Northrop-Grumman (Baltimore, Maryland)

and Raytheon (Plano, Texas) which produces the AN/APG-77(V)1 radar for the F-22A. Figure 4.1 shows the timeline of visits and meetings among RAND, the SPO, the prime contractors, and four major subcontractors. Figure 4.2 presents the percentage breakdown of the savings initiatives proposed by the prime contractors and the major subcontractors.

At each site, we evaluated the proposed savings initiatives and the associated supporting analysis provided with each, including required investments, to substantiate how realistic and reasonable they were. To do this, we examined the production area where the savings would occur and compared the single-year and multiyear cost estimates to validate the computed savings. The feasibility of each initiative was then evaluated and, if accepted, was traced into the specific task sheet and negotiated value (by contract line item number) in the multiyear contract. The last step was undertaken during the last week of May 2007.

To help explain sources for future multiyear contract considerations, we organized the initiatives into six categories, as explained below.[1] Unlike previous reports that documented savings and general reasons and sources for the savings, our report sought to explain the relationship between these savings initiatives and the reduced values in the multiyear contract. During our analysis, we used a category called "management challenge" as a temporary category for those initiatives that lacked sufficient justification to be included as savings. We asked the contractors for more details or data on those initiatives during the five-month evaluation process. In the final analysis, initiatives in the management challenge cate-

[1] As part of the contract award process, the USAF asked the prime contractors to submit two proposals each. One addressed a single-year contract for 20 aircraft (Lot 7) in FY 2007 and the other addressed a multiyear contract for 60 aircraft (Lots 7–9) in FYs 2007–2009. Since the multiyear award was contingent upon a certification by the Secretary of Defense, both the single and multiyear options were fully negotiated and prepared for award, because neither the USAF nor the contractors knew whether the multiyear would be approved. These negotiations included negotiated prices between the prime contractors and their subcontractors and suppliers. As a result, the negotiated single-year Lot 7 values were one source of valuable data in establishing savings between the two options.

Figure 4.1
Project Timeline and Meetings on Savings Initiatives

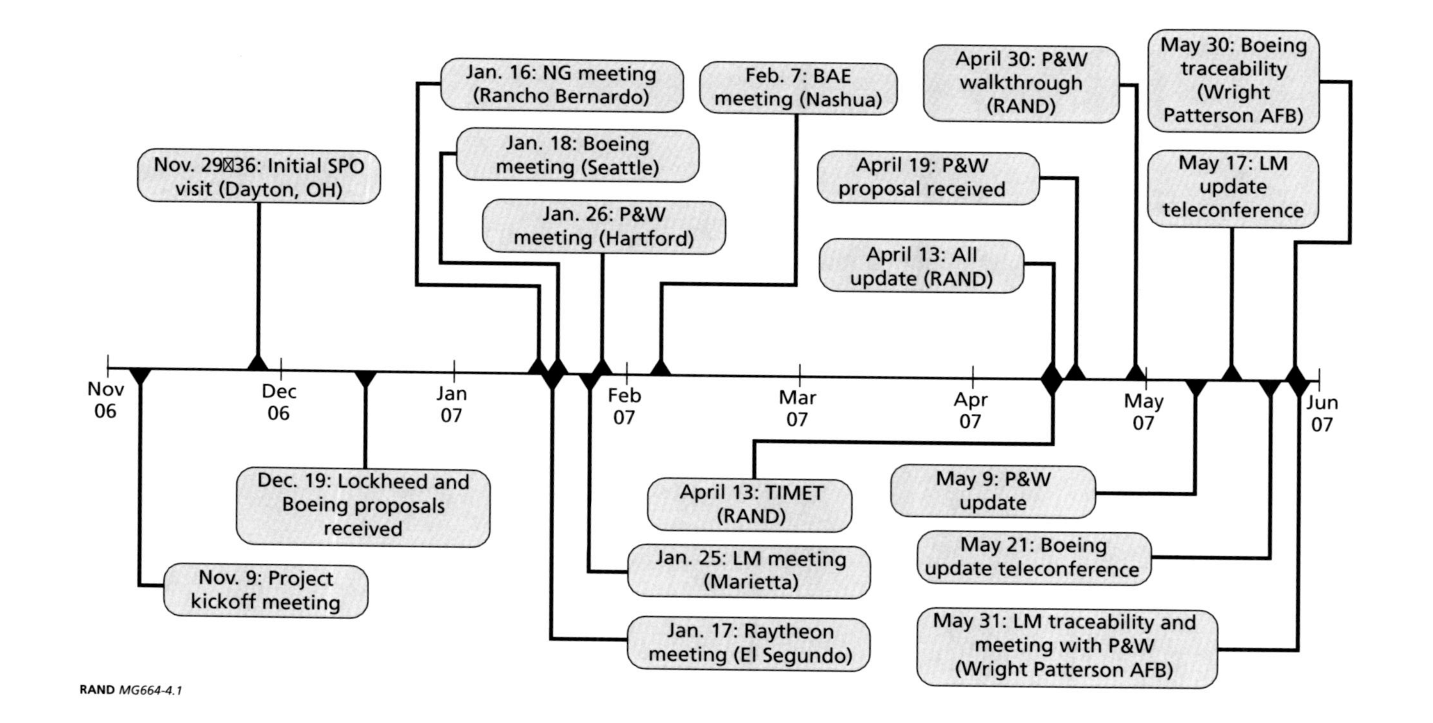

RAND *MG664-4.1*

Figure 4.2
Percentage of Savings Initiatives by Dollar Value as Proposed by Various Contractors

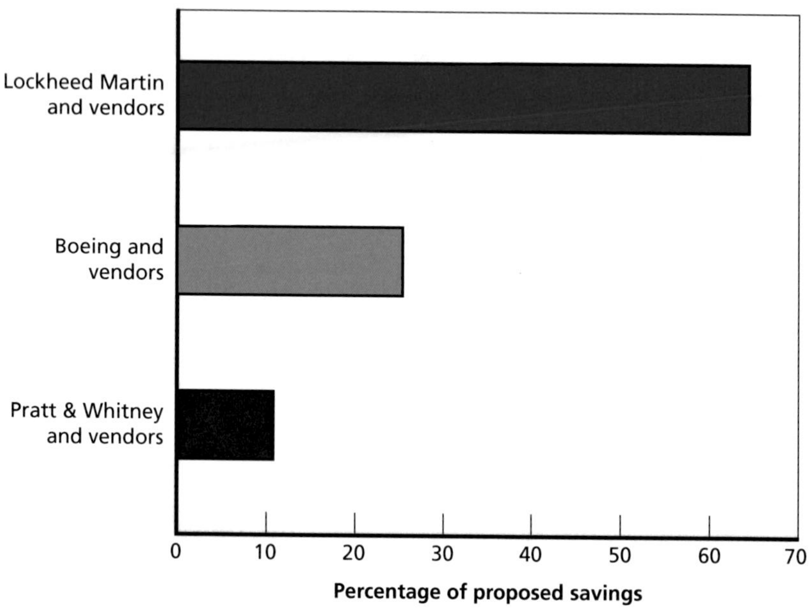

RAND *MG664-4.2*

gory were either justified and moved to another category or labeled unsubstantiated and not credited as savings. Of the $311 million in savings proposed by the contractors, $15 million did not meet the above criteria. All $296 million in substantiated savings initiatives were reflected in the multiyear contract negotiated between the contractors and the USAF.

Quantitative Factors in Multiyear Contracts

Many factors favor savings under a multiyear contracting arrangement. Appendix B contains a listing of factors from previous studies of multiyear contracts by source and the year in which the contract was awarded or the study written. These studies include those by the GAO,

IDA, and RAND. A number of factors are categorized differently or overlap one another in the various studies, so a universal categorization scheme using all (or even a majority of) the studies was not feasible. For our purposes, we developed six categories that could best explain the source of the savings from a multiyear contract. These six categories are explained in the following paragraphs. Table 4.1 shows the contribution of each category to the total claimed savings of $311 million and the RAND-substantiated savings of $296 million.

Alternative Sourcing

A multiyear contract offers the certainty of larger quantities and a longer production run. These can make a stronger economic case for the prime or larger suppliers to spend the time and energy to seek out alternative sources of supply that could result in savings. These savings could result from more efficient production processes, from alternative vendors, or just a general lowering of prices from the vendors due to real or perceived competition. The F-22A program, however, has relatively small numbers of aircraft yet to be produced compared with other aircraft multiyear programs, so the time and cost of further competition at the vendor level would not be as productive as in a program with significant quantities remaining. The F-22A program has three initiatives: down-select from two vendors to one, change the supplier

Table 4.1
Contractor-Claimed and RAND-Substantiated Savings by Category

Savings Categories	Contractor-Claimed Savings (TY $millions)	Substantiated Savings (TY $millions)	Percentage of Substantiated Savings
Alternative sourcing	11	11	4
Production build-out or acceleration	48	48	16
Buyout of parts and materials	130	130	44
Proposal preparation	25	25	8
Support labor	82	82	28
Management challenge (unsubstantiated)	15	0	0
Total	311	296	100

source, or move some of the workload currently performed in-house to a supplier. In the first case, two sources of a major airframe section are being reduced to one to allow use of the less-expensive approach on all remaining aircraft, with estimated net savings of about $8 million during the multiyear contract. In the second case, about 60 composite parts will be outsourced from the LM Marietta plant to suppliers, with a net savings of almost $3 million. These three initiatives are estimated to produce net savings of $11 million, or about 4 percent of the total substantiated multiyear savings.

Production Build-Out/Production Acceleration

In cases where annual setup costs are significant and production quantities are well below efficient steady-state production rates, initiatives to build all remaining ship sets of parts or subassemblies could result in direct labor and support labor savings. In a build-out or acceleration, production rates are increased so that use of touch labor can be increased on the production line, the line can then be terminated at the completion of the production run, and support activities can be significantly reduced or eliminated at that point. This makes sense especially on production lines where the F-22A constitutes all of the output.

We categorized initiatives into production build-out or acceleration; in both initiatives key subcontractors producing more sophisticated parts or subassemblies were involved and the production strategy was evident, as opposed to the less-complex parts in the following buyout category. Examples are the build-out of the high-density multilayer interconnect (HDMI) and the common backplane assembly (CBA), both associated with Raytheon. The $48 million in savings in this category of initiatives constituted about 16 percent of the total multiyear savings. Only three initiatives required EOQ funding. EOQ funding of $8 million produced about $2 million in savings, for a 24 percent return on investment.

Buyout of Parts and Materials

In general, the parts and materials "bought out" are less sophisticated parts and assemblies than those "built out." A total of 52 initiatives were identified, 44 of which utilized EOQ funding, yielding an average return on investment of about 36 percent. The highest return for one initiative was over 400 percent and the lowest was 13 percent. The initiatives were selected by the prime contractors based on their individual return on investments (ROIs). The prime contractors and major subcontractors noted that they had developed many more initiatives than could have been implemented had additional EOQ dollars been available. They felt that another $250 to $300 million could have been used and still yielded ROIs of over 10 percent. As was the case with the previous category, the majority of these savings occur during Lots 8 and 9 since many long-lead parts and subassemblies have already been ordered for Lot 7 and because production can be completed sometime in Lot 8 or early in Lot 9. In some cases, orders placed for Lot 7 can be merged with Lots 8 and 9 under the multiyear contract to achieve savings for all three lots. The savings from this category represented about 43 percent of the overall multiyear savings, or about $130 million.

Proposal Preparation Savings

One clear-cut area where multiyear contracts eliminate workload concerns the annual activities related to proposal preparation at all levels (prime contractors, subcontractors, and certain suppliers), analysis of subcontractor and supplier proposals by the prime contractors, pre-award fact-finding, and contract negotiations. Because of the many requirements of the Federal Acquisition Regulation for contract award, these activities can take anywhere from six months to a year and result in proposals that are literally hundreds of pages thick. Other supporting documentation and analyses of subcontractor and supplier proposals to the prime are also required, including the requirement for the Truth-in-Negotiations Act (TINA)–certified cost data for prime contractors and major subcontractors and suppliers.[2] With the awarding of the multiyear contract, no proposals from the prime contractors or

[2] Federal Acquisition Regulation Part 15.403 requires certified cost and pricing data to be submitted by contractors for any anticipated award of $550,000 or more where cost data are required as part of the award process. In the case of a multiyear contract, this threshold would apply to almost the entire contract award, or three lots of parts or materials for the F-22A MYP.

major subcontractors and suppliers are required for Lots 8 and 9, resulting in savings of at least $25 million. Proposal activities begin in the fiscal year before contract award and are generally charged to the prior year's funding (depending on company policy), so most of the savings actually occur in Lots 7 and 8. This figure primarily captures the direct cost of these activities (where people specifically charge their time to the F-22A activities in the contractors' cost accounting systems), but does not capture the majority of the indirect activities accounted for in overhead rates and factors—nor does it capture the savings at the smaller suppliers and vendors. The scope of the project did not allow collecting savings data in any more detail than those reported by the prime contractors and major subcontractors, but we would estimate the savings would be significantly greater, both in the overhead area as well as at the smaller companies, than the directly reported cost savings.[3] This category constituted about 8 percent of the overall multiyear savings estimate.

Support Labor Savings

As with the proposal preparation activities, there is a myriad of support activities—production planning, engineering, tooling support, supplier management, financial analysis and reporting, cost estimating and pricing, contract administration, and other non-touch labor. Most of these indirect activities are based on historical factors and are applied to cost estimates and contract payments using direct labor hours or material dollars as the basis for allocation. Thus, a reduction in production costs or an acceleration of supplier production and deliveries can decrease indirect labor costs. The estimate for the reduction in this category is approximately $82 million, or about 28 percent of the multiyear savings.

---

[3] Lorell and Graser, 2001, page 49. Proposal preparation costs were estimated to be about 1 percent of the contract value. This would equate to about $60 million in savings for Lots 8 and 9 of the F-22A MYP.

Management Challenge

This category was to include initiatives that we were unable to substantiate. Direct quantification of reduced costs for the final category was the most difficult, and several items could not be directly related to a specific reduction in contractor prices. Figure 4.3 shows the evolution of the savings initiatives over the analysis period. As previously stated, we used the management challenge category as a temporary holding category while we did more analysis and gathered more data. Ultimately, each of these initiatives was either substantiated and moved to another category or rejected and eliminated from the savings total.

In some cases, some of these initiatives may provide savings to the USAF, perhaps on other contracts, but they could not be counted with the ground rules we established for the MYP program. For example, once the prime contractors negotiated prices for parts or materials for the multiyear contract, these same prices would also be used for purchases of spare parts under sustainment activities, funded under a different contract.

Qualitative Factors

The focus of arguments for or against multiyear contracts is often on the size of the projected, quantified savings, but several advantages to both DoD and the contractors from a multiyear contract may not be quantifiable. Under proper circumstances, they may help make the case for awarding a multiyear contract. First, the certainty of production under a multiyear contract allows for a more level factory loading, better production scheduling, and longer-range material procurement planning. The stability of the business base provided by a multiyear contract also allows for more predictable forward pricing rates and overhead factors and a better ability to forecast a contractor's rates and factors on other DoD programs produced in the same plant or business area. This pre-

Figure 4.3
Evolution of Contractor-Proposed Savings

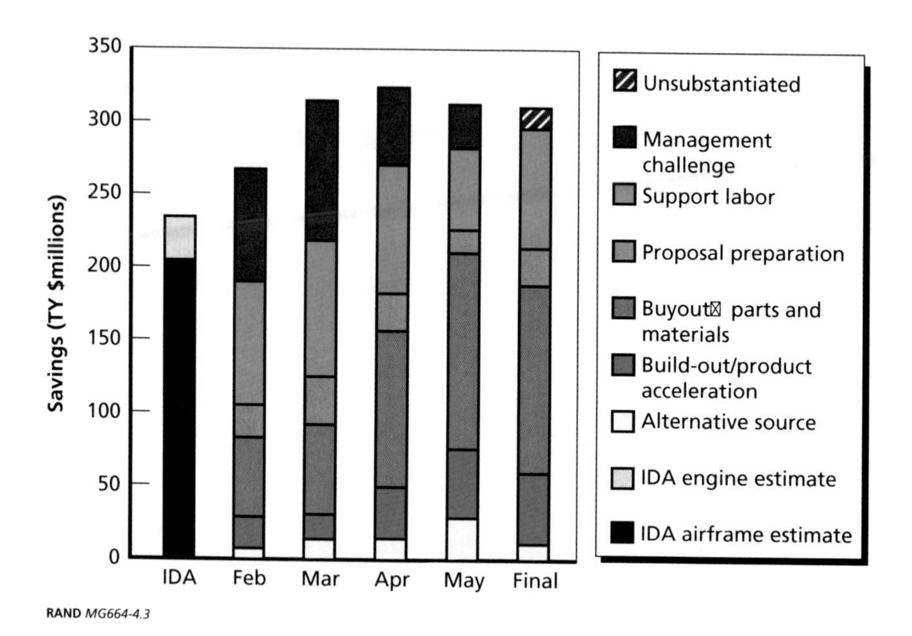

RAND *MG664-4.3*

dictability can be used to develop budgets for all programs involved during the PPBS and congressional activities.

As a mirror image of savings on the contractor side, a multiyear contract also reduces the government workload by eliminating the need for RFPs for later lots, government fact-finding and negotiations on later lots, and the number of active contracts at any one time, thereby simplifying contract administration. For example, one contract closeout exercise can be conducted at the end of the multiyear contract, rather than performing one for each annual lot. Since this study was focused on savings generated by the multiyear contract, we did not attempt to quantify the dollar savings to the USAF for these activities.

Areas Where Little or No Savings Occur Under MYP

Despite the quantifiable savings and qualitative advantages, many areas remain basically the same under single-year and multiyear contracting. Many of these are due to insufficient funding in the FY 2007 budget and beyond to accelerate the entire F-22A program. For example, production is still planned for 20 aircraft per lot, so major assembly and final assembly and checkout activities (such as touch labor and associated overhead) will be basically the same under the multiyear contract. In addition, with the limited EOQ funding available, fabrication of large assemblies will be nearly identical under the multiyear and single-year contracts. Since the production program continues well into FY 2011, program management activities continue under either contract type, as will tail-up costs and financial reporting. Other activities that are directly identifiable with production or individual hardware items, such as materials consumed in production, would also be the same in either contracting case. In addition, some level-of-effort activities (such as much of the plant overhead and general and administrative costs) would be the same under either case.

Summary

This chapter discussed the quantifiable and nonquantifiable aspects of the F-22A multiyear contract, the advantages of each, and the portion each category contributed to the overall substantiated savings of $296 million for the multiyear contract, which accounts for over 70 percent of our estimated savings of $411 million. It also noted that certain activities remain basically the same under either contracting case since the constrained budget prevents the overall F-22A program from being accelerated to a more efficient annual rate.

CHAPTER FIVE

A Look at the History of Aircraft Multiyear Contracts

In this chapter we provide a historical context for aircraft and aircraft-related multiyear procurements dating back to 1982, as required by the congressional language authorizing this research study.

Introduction

This chapter provides an overview of all 14 major multiyear procurement contracts, for fixed-wing military aircraft from 1982 to the present, that RAND identified. The chapter is divided into five sections.

The first section provides a brief overview of the key elements in past MYP contracts that cost estimators and other expert observers believe affect the magnitude of estimated cost savings as a percentage of contract value and that complicate comparisons among varying historical MYP contracts. These factors were identified by reviewing the open literature on past MYP contracts, with a heavy emphasis on GAO reports, and through interviews with DoD cost analysts and acquisition experts. These categories of characteristics are in addition to the statutory characteristics that all multiyear procurements must possess as delineated in 10 USC, Section 2306b.[1] We also include a brief men-

[1] As noted in Chapter Two, 10 USC 2306b requires that (1) the multiyear contract will result in "substantial savings," (2) the requirement for the end item is stable, (3) procurement funding is budgeted and stable, (4) the end item design is stable, (5) the cost savings or avoid-

tion of the key areas historically identified by DoD and GAO cost estimators as the principal sources of cost savings on historical MYP contracts.

The second section quickly reviews seven older MYP contracts from 1982 through 1995. These programs are treated with brevity because the regulatory, legal, and industrial environments were less like today's, and because of the general lack of detailed qualitative and quantitative information due to their age. The programs included in this section are F-16 MYP I, II, and III; KC-10; B-1; C-2; and AV-8B Harrier.[2]

The third section reviews several more recent MYP contracts from the post-1995 period in somewhat greater detail. These programs generally have more qualitative and quantitative information available, and often took place in a regulatory, legal, and industrial environment that is more similar to the present. The programs included in this section are C-17 MYP I and II; E-2C MYP I and II; C-130J; and F/A-18E/F MYP I and II.[3] Some of the detailed information about the past contracts appears in Appendix E.

The F/A-18E/F is the only recent fighter aircraft procured through an MYP contract. For this reason, the fourth section delves into this MYP program in greater detail. Some quantitative analysis of available cost data is presented. First we attempt to validate the originally estimated savings used for the MYP program justification. Then we try to normalize the F/A-18E/F data to make key program characteristics more comparable to the proposed F-22A MYP for purposes of analysis. That analysis can be found in Appendix A. However, even for the F/A-18E/F program, the data availability and quality limit the scope and applicability of the analysis.

---

ance estimates are "realistic," and (6) use of a multiyear contract will promote the national security of the United States.

[2] A second AV-8B MYP contract followed the first, but it is not reviewed here. This is because the AV-8B MYP II is an unusual program involving remanufacturing and updating existing aircraft. In addition, a significant amount of the work was undertaken by government depots rather than by private contractors.

[3] Further detail is provided on these and more recent MYP aircraft programs in Appendix E.

Finally, the last section provides a brief summary and findings based on our review of historical MYP fixed-wing military aircraft contracts.

Varying Key Program Attributes of Historical MYP Contracts

Appendix B reviews the key sources of savings on MYP contracts as identified by multiple studies and other sources. In contrast, the box on the next page identifies a variety of key program attributes of historical MYP contracts that complicate cross-program comparisons and that can affect the expected percentage of savings compared with SYP contracts possible in any given MYP program. We identified these specific attributes based on our review of the acquisition literature and through extensive interviews with government and industry acquisition officials. These attributes are listed and briefly discussed below. In most cases, however, few or no data are available to support a quantitative analysis of how and to what extent these attributes affect MYP savings percentage estimates for any given program, or the degree to which variation in these attributes must be corrected when comparing one historical MYP contract savings percentage estimate with another.

Later in this chapter, we report the results of some simple correlation analysis we conducted on the seven most recent programs to see if any statistically significant correlations exist between any of these characteristics and the estimated level of savings percentage for an MYP.

Validation of Claimed Savings

It is extremely important to emphasize that all the savings for historical MYP contracts reported here are estimates developed before the multiyear program was actually approved and implemented. Claimed savings for multiyear programs are derived by comparing pre-program estimates of the costs of a multiyear procurement with estimates of the

Key Program Attributes of Historical MYP Contracts

Presence and Scale of EOQ Funding

EOQ funding assists the prime contractor in purchasing items from vendors in quantities greater than those that are required for production in any given fiscal year. In theory, this enables the prime contractor to realize price reductions from vendors by buying items in larger, more economically efficient quantities. Historical MYPs vary in the amount of EOQ funding as a percentage of the prime contract that prime contractors received, and some programs received no EOQ funding at all. Furthermore, some historical programs were inherently more suited to take advantage of EOQ funding than others because of technical aspects of the products and other factors.

Phasing of EOQ Funding

The timing and phasing of EOQ funding could also in principle have a significant effect. All things equal, the availability of significant EOQ funding early in a program and well in advance of the first MYP annual lot would likely have a greater beneficial effect than in the opposite situation.

Annual and Total Procurement Numbers

It has been argued—although not proven with data—that MYP programs with large annual and total procurement numbers provide a much greater opportunity for efficiencies of scale in the use of EOQ and Cost Reduction Initiative (CRI) funding than programs with smaller quantities.

Duration of MYP Contract

There is some variation in the length of historical MYP contracts. In a manner similar to the annual and total procurement numbers, longer MYP contracts are seen by some as providing greater opportunities for EOQ purchasing and implementation of manufacturing efficiencies and other nonrecurring CRIs that result in price reductions.

Presence and Scale of CRI

Not all historical MYPs include CRI funding. While some observers have argued that CRIs are not necessarily uniquely associated with MYP contracts, the presence and scale of CRI funding during MYP contracts are likely to have significant effects on program outcomes.

Business Base and Industry Economic Environment

Trends in the business base of the prime contractor, as well as in the aerospace industry as a whole and on lower tiers during any given MYP, can complicate cross-comparisons and comparisons between SYPs and MYPs for the same system, because overhead and other indirect costs and negotiating leverage with suppliers and vendors may both be affected.

Production Rate Changes

Significant production rate changes when comparing SYPs with MYPs for the same system can complicate comparisons.

Maturity of Production Program

The phase in the overall anticipated production life of the program in which the MYP takes place may affect outcomes. Some have argued that MYPs that take place at or near the anticipated conclusion of a program may experience smaller savings percentages than those that take place during earlier phases of the production cycle, for at least two reasons. First, a less mature production program may provide more CRI opportunities to make production more efficient. Second, MYP programs that take place near the anticipated end of the production life may appear less attractive to vendors and thus elicit lower price savings and discounts, even with EOQ funding.

Prospects for Foreign Sales

Obviously, MYP programs with significant foreign sales prospects enjoy the benefits of increased production numbers and rates.

total costs of several sequential hypothetical single-year procurements covering the same period. Some of the estimates are based on better data and methodologies than others, but all are estimates, in almost all cases carried out before MYP approval. In our survey of the literature and our search through DoD and company acquisition documents, we found very few examples of serious and methodologically credible attempts to validate claimed savings and savings percentages after the fact once programs had been completed.

There are two major reasons for this. First, once programs are approved and implemented, various important program assumptions on which the original savings estimates were based often changed, and normalizing for these changes is difficult. Secondly, the government neither collects nor saves the data necessary to conduct a detailed analysis of actual cost savings. And of course, the single-year procurements used for comparison always by definition will be hypothetical estimates.

For example, one of the most extensive and thorough independent attempts to validate claimed savings on an MYP contract was undertaken by the GAO in 1985 at the request of Congress. The GAO attempted to validate the claimed savings on the F-16 MYP I contract for FYs 1982–1985. While the GAO found after extensive analysis of vendor prices that savings in that area were likely realized, it concluded that "we could not determine if total savings projected by the Air Force were achieved."[4] This was because key program attributes changed after the original estimate was prepared and the program was approved. The GAO also concluded, "[t]here are no comparable multiyear and annual cost estimates available for us to confirm whether . . . savings estimated by the Air Force . . . were achieved."[5] Similar situations arise on nearly all historical MYP contracts. A few other attempts have been made to validate estimated MYP contract savings after the fact, but none that we are aware of have produced definitive findings.

Thus, it must be remembered that all savings reported below on historical MYP contracts are merely estimates presented as justifications to Congress before the final negotiation and approval of the MYP contract. We have not attempted to validate those savings due to limitations in data availability and quality. The one partial exception is the F/A-18E/F MYP I program discussed later in this chapter. However, as that discussion shows, even with more data available it is extremely difficult to arrive at definitive conclusions regarding actual savings and comparability to the proposed F-22A MYP.

Finally, it is also important to note that all savings estimates, as well as other supporting budgetary and cost data, are presented in then-year dollars. This is because, to the best of our knowledge, all official current and historical MYP savings estimates, as well as accompanying cost data of all types, are reported only in TY dollars. Justifications for multiyear procurements are part of the DoD budgeting process. The documents are incorporated into the service's budget justification books and, as is normal in all budget-planning documents, are shown only in TY dollars. Adjusting historical estimates based on TY dollars to some arbitrary base-year cost poses many methodological challenges, because the current standard deflators can differ significantly from those used by cost analysts on past programs to develop and project their original estimates out into the future in TY dollars. Adjusting historical estimates using current deflators based on history thus could significantly distort the original estimate, which might legitimately have been based on different assumptions and projections about the future. Therefore,

---

[4] GAO/NSIAD-86-38, February 1986, p. 2.

[5] GAO/NSIAD-86-38, p. 2.

all historical savings estimates and most other cost and budgetary data shown here remain in the original TY dollar terms. Except possibly for a few programs in the early 1980s when inflation was high—programs that we treat only in passing—this approach should not cause significant distortions and is analytically consistent.

Military Fixed-Wing MYP Contracts, 1982–1995

As noted previously, we identified seven relevant fixed-wing combat aircraft MYP programs between 1982 and 1995, including three MYP contracts involving the General Dynamics (now Lockheed) F-16 fighter aircraft. These programs were all launched nearly two decades ago during the Cold War era and took place in dramatically different industrial base conditions and acquisition environments. In addition, given the length of time since the programs' inception, it is far more difficult to find and verify detailed data and reliable information from actual program participants. Therefore, we report here only the most basic information about these programs, with the partial exception of the first F-16 MYP program, which was extensively analyzed by the GAO in 1986. More detail is provided on the seven more recent fixed-wing military aircraft programs launched after 1995, as discussed in the next section.

F-16 MYP I

By the time Congress approved the F-16 MYP I, more than 500 F-16s had already been delivered to the Air Force and foreign customers. A significant amount of information is available on the F-16 MYP I, mainly because the GAO undertook an extensive study at the request of Congress to validate the claimed savings estimates. According to the GAO, in the original justification submitted to Congress in October 1981, the Air Force estimated a total procurement over a four-year multiyear contract (FYs 1982–1985) of 480 F-16 aircraft for a savings of TY $246 million compared with SYP, or 7.7 percent savings. The MYP savings were estimated by comparing the estimated total SYP contract cost of $3.184 billion against the estimated multiyear price of $2.938 billion.[6]

However, the MYP estimate went through several major changes from the time General Dynamics submitted its original MYP contract proposal, compared with SYP proposals submitted in March 1981. The Air Force adjusted the original contractor estimates after deciding to procure a large number of significantly upgraded F-16C/D variants during the MYP I. Furthermore, some additional Air Force aircraft were later added to the buy, and the decision was made to produce all aircraft cooperatively with European program partner countries (The Netherlands, Belgium, Denmark, and Norway). These four countries also decided to procure an additional 146 airframes during MYP I, but these were priced under a separate contract. Many of these changes, as well as different assumptions regarding inflation, were reflected in the formal justification submitted to Congress in October 1981. Changes in the program also took place after approval of the MYP, but the contractor was never required to adjust and update its original SYP proposals from March 1981 for a comparison to reflect these changes.

According to the contractor and the Air Force estimates, as reported by the GAO, EOQ funding and advanced purchase of subsystems and materials made up by far the largest source of cost savings. The Air Force estimated that economic orders of subsystems would account for about 42 percent of savings, and general material procurement would account for another 32 percent. The GAO calculated that about 46 percent of subsystem savings was due to combining the foreign orders with the Air Force procurement.[7]

Published discussions of this program often use an Air Force estimate based on different inflation assumptions that were prepared prior to the congressional justification package. This estimate projected a savings of TY $350 million with an estimated multiyear contract of $2.986 billion, for a total contract savings of 10.5 percent. Some sources use this as the definitive estimate and combine the foreign orders and addi-

---

[6] GAO/NSIAD-86-38, February 1986.

[7] GAO/NSIAD-86-38, February 1986, p. 3.

tional Air Force orders, for a total of 534 aircraft for MYP I, retaining the estimate of overall multiyear savings of 10.5 percent.[8]

However, to maintain consistency with the other historical MYP programs presented here, we use the estimate presented to Congress in the formal MYP justification.

Other MYP Programs, 1982–1995

Because of the relative dearth of detailed information available for the six other major fixed-wing combat aircraft MYPs during this period, we decided to summarize the most important data about these programs in a single table (Table 5.1). No further details of these programs are included here, although we present a few general observations. A quick review of Table 5.1 shows that the estimated contract savings on the seven fixed-wing military aircraft MYP programs between 1982 and 1995 varied from 5.7 percent on the F-16 MYP III to 17.7 percent on the KC-10 MYP program. Excluding the KC-10 MYP program, which heavily leveraged commercial technology based on the McDonnell Douglas DC-10 wide-body commercial transport aircraft, the savings estimates ranged from 5.7 percent to 11.9 percent. This is very close to the range of savings estimates experienced on the post-1995 MYP programs, as shown in the next section. Five of the programs were relatively large in budgetary terms compared with the other two, whereas only three (F-16 I, II, and III) entailed large production numbers. Estimated contract savings percentages were significantly larger for the programs with smaller procurement numbers than for those with larger procurement numbers, although there appears to be no correlation between program budgetary size, procurement numbers, and estimates of contract savings percentages. Too little is known about the proposed EOQ and CRI funding to attempt to draw any correlations, although the program with the highest contract savings percentage estimate (the KC-10) also had a very high contract percentage level of EOQ funding (15.8 percent of the contract).

---

[8] These numbers are reported in an internal F-16 SPO memorandum by an unknown author dated July 1999. They are also reported in Air Force Materiel Command, 1996.

Table 5.1
Fixed-Wing MYP Contract Cost Savings Estimates, 1982–1995

MYP Program	Estimated MYP Contract Value (airframe) (TY $billions)	Estimated SYP Contract Value (airframe) (TY $billions)	Estimated Contract Savings (TY $billions & %)	Total Quantity	Time Frame (Total & FYs)	EOQ Funding (TY $millions & %)	MYP AP Funding (TY $millions)	SYP AP Funding (TY $millions)
F-16 I	2.9	3.2	0.25 7.7%	480	4 yrs. 82–85	Unk.	Unk.	Unk.
F-16 II	3.9	4.2	0.36 8.4%	780	4 yrs. 86–89	82 2%	720	639
F-16 III	4.3	4.6	0.26 5.7%	630	4 yrs. 90–93	Unk.	Unk.	Unk.
B1-B	10.6	11.8	1.19 10.0%	92	4 yrs. 83–86	908 8.5%	3,816	2,908
C-2	0.7	0.7	0.06 7.9%	39	5 yrs. 83–87	115 17%	197	82
KC-10	2.8	3.4	0.6 17.7%	44	5 yrs. 83–87	441 15.8%	44	Unk.
AV-8B	0.9	1.0	0.12 11.9%	72	3 yrs. 89–91	Unk.	Unk.	Unk.

NOTE: Unk. = unknown.

Another interesting aspect of the early programs is that they clearly represented a major but short-lived surge of interest in MYP programs following congressional legislation in 1982 that loosened the restrictions on establishing multiyear procurement programs and providing EOQ funding.[9] Four of the seven MYP programs included in this group were launched in the narrow FY 1982–1983 time frame. Of the three remaining programs, two (F-16 II and III) were follow-ons to one of the original programs funded after the 1982 legislation, the F-16 MYP I. Thus, only one of the MYP programs (the AV-8B Harrier) was launched as a totally new effort after the initial 1982 surge in interest in MYP programs, and it was one of the smallest programs, with a total contract value of under TY $1 billion. We could not determine what led to this rapid drop-off in interest. However, it is well known that DoD submitted numerous potential MYP candidates for congressional consideration every year during FYs 1982 through 1995. We can only assume that Congress felt uncomfortable with approving additional MYP programs after the initial surge of approvals for FYs 1982–1983.

Military Fixed-Wing MYP Contracts Since 1995

This section reviews more-recent MYP contracts in somewhat greater detail than the earlier contracts. Those contracts that have one or more key characteristics similar to the proposed F-22A MYP contract are discussed in somewhat greater depth.

---

[9] In April 1981, then Deputy Secretary of Defense Frank C. Carlucci presented 32 procurement reform initiatives intended to reduce weapon system procurement costs. One of these initiatives urged Congress to reduce the statutory and regulatory restrictions on the use of MYP programs, which had been instituted in the 1970s. Congress removed the restrictions in time for application of MYP funding in the FY 1982 defense budget. Advocates of increased use of MYP contracting expected this reform to save 10–20 percent on contract costs. See Foelber, 1982.

C-17 MYP I

The Boeing (formerly McDonnell Douglas)[10] C-17 Globemaster III is the USAF's premier strategic airlifter. In May 1996, the USAF and McDonnell Douglas signed a $14.2 billion MYP contract for 80 aircraft to be procured over seven years (FYs 1997–2003). This was the largest and longest multiyear contract ever negotiated up to that point. According to the original proposal, the contractor and the Air Force estimated that the MYP contract with McDonnell Douglas for 80 aircraft would save 5.5 percent or about $900 million (all in TY$), compared with a series of SYP contracts over a comparable period for the same number of aircraft.[11] At the same time, the Air Force also signed a multiyear contract with Pratt & Whitney for the commercial derivative engine (CDE) program for procurement of the F117 engine to power the C-17. This contract was valued at $1.6 billion; the Air Force estimated the multiyear procurement program for the engine saved $88 million, or 5.5 percent over SYP.[12]

---

[10] Boeing announced the planned acquisition of McDonnell Douglas for $13.3 billion in December 1996.

[11] See Congressional Research Service, 2000, p. 3. There is some uncertainty from the available data as to whether the final savings estimate shown above for the aircraft included multiyear savings on the engine program. This is because, unlike all other fixed-wing multiyear programs after 1995, no formal multiyear justification for Congress was published at the time of program initiation, because the program was proposed outside the normal budget cycle. According to a contemporary GAO report published before the contract negotiation with McDonnell, the original total program savings estimate was for 5 percent, or TY $896 million. This account claims the estimated savings came primarily from two sources: the airframe contract and the engine contract. According to this account, McDonnell Douglas reduced its contract price for 80 aircraft by about $760 million, or 5 percent. The USAF expected to realize a 6 percent savings or approximately $122 million on the engine contract with Pratt & Whitney. The Air Force estimated that the remaining savings of about $14 million would come from other sources. See GAO/T-NSIAD-96-137, March 28, 1996, p. 6. However, other sources seem to indicate that the $900 million estimate of savings in the final contract with McDonnell did not include the engines, e.g., see GlobalSecurity.org, 2005.

[12] Beginning with the Lot 4 buy in November 1992, the government took over procuremet of the F117 engine and provided it to McDonnell Douglas as government furnished equipment (GFE). See DoD, 1996.

C-17 MYP II

The C-17 MYP II program, covering the procurement of 60 aircraft over five fiscal years (FYs 2003–2007), claimed $1.3 billion cost savings over an estimated annual procurement cost of $12.8 billion, or an overall 10 percent cost savings for the airframe and engine contracts. The Air Force estimated a 10.8 percent cost savings on the airframe procurement contract with Boeing over annual procurement contracts (or $1.211 billion and a 5.7 percent cost savings). More specifically, the Air Force estimated the annual cost of an SYP program at $12.805 billion versus $11.503 billion for an MYP program.[13]

E-2C MYP I

The E-2C Hawkeye is a U.S. Navy carrier-based tactical airborne warning and control system platform. The first E-2C MYP contract covered FYs 1999–2003. It was a single five-year FFP contract for the airframe only. The original Navy MYP justification (February 1998) showed 8.3 percent total airframe contract savings, or $106.5 million over annual contracts with the same quantity profile, with the total MYP airframe contract price originally estimated at $1,181.3 million. The total procured quantity was very low: 21 aircraft. There was a relatively large amount of EOQ funding for material for 21 ship sets of "detail parts," and contracting for 21 ship sets of "Prime Mission Equipment" was provided during one lot buy in FY 1999. Indeed, over one-third of the contract value was EOQ funding. GFE included engines and the Joint Tactical Information Distribution System (JTIDS). CFE included the radar, Passive Detection System (PDS), rotodome, landing gear, Identification Friend or Foe (IFF), and other equipment.

In April 1999, Northrop-Grumman was awarded a $1.3 billion five-year MYP contract covering 22 Hawkeye 2000 (upgraded E-2Cs), which included 21 for the U.S. Navy and one for the French Navy. Later, two more foreign military sales aircraft were added.

---

[13] DoD, 2003a, pp. 2–3.

E-2C MYP II

The E-2C MYP II contract in many respects does not constitute a true MYP, according to the Naval Air Systems Command (NAVAIR). It is the smallest fixed-wing military MYP contract since 1995. Significant savings were not claimed to be the primary motivation for the MYP contract. Rather it was an attempt to fill a production gap and keep the production line warm between the end of E-2C production and the beginning of LRIP for the significantly upgraded E-2 Advanced Hawkeye as it was called at the time. However, the Navy formal justification did claim a cost savings over annual procurement of 7.2 percent for the airframe and engine procurements. The E-2C MYP II consisted of two four-year fixed-price contracts, one for the engines and one for the aircraft, covering FYs 2004–2007. The entire procurement consisted of four E-2C aircraft and four TE-2C aircraft and 16 engines. The total MYP procurement contract price (airframe and engine) was estimated in the justification at $788.6 million, compared with an annual total contract price estimated at $850.0 million, for an estimated cost savings of $61.4 million for the airframe and engine contracts.

CC-130J (USAF) and KC-130J (USMC) MYP

In March 2003, the Air Force awarded Lockheed Martin a $4.05 billion six-year joint Air Force–Marine Corps MYP contract for procurement of 60 CC-130J and KC-130J Super Hercules aircraft from FYs 2003 through 2008.[14] This includes 40 CC-130Js, a stretched version of the C-130J tactical airlifter being procured by the U.S. Air Force, and 20 KC-130Js, an aerial tanker/transport version of the C-130J procured by the U.S. Marine Corps. As of mid-2006, total annual deliveries of both types combined were scheduled as follows: 4, 4, 15, 13, 13, 11.[15] MYP total airframe contract savings compared with SYP were originally estimated at 10.9 percent, or $513.07 million.

---

[14] DoD, Office of Inspector General, 2006, p. 4.

[15] The original MYP justification envisioned a 62-aircraft procurement with total annual procurement beginning in FY 2003 as follows: 12, 13, 12, 13, and 12. Forty-two USAF CC-130Js are shown on the Contract Funding Plan as a five-year MYP, with four procured in FY 2004 and the remaining 38 procured from FYs 2005 to 2008. KC-130J Marine aircraft are

This is a fixed-price contract with production/quantity rate change adjustment factors. Originally it was a commercial item FAR Part 12 Price-Based Acquisition (firm fixed price with economic price adjustments, or FFP + EPA) contract, with no formal cost and price reporting. This contract is currently being restructured as a traditional FAR Part 15 contract.

F/A-18E/F MYP I

In June 2000, the U.S. Navy launched full-rate production of the F/A-18E/F Super Hornet fighter by signing a five-year MYP contract (FYs 2000 to 2004) with Boeing for $8.9 billion covering the procurement of 222 aircraft (later reduced to 210). The original justification documentation claimed a cost savings over annual contracts of TY $706 million, or 7.4 percent on the total airframe contract price with Boeing. The contract was a fixed price incentive (FPI)–type contract, with a 70:30 split.

An additional five-year MYP engine contract covering FYs 2002 through 2006 was proposed in 2002, with estimated savings of 2.8 percent. In July 2002, the U.S. Navy awarded GE a $1.9 billion five-year MYP contract for 480 F414 engines, devices, and spare modules.

According the F/A-18E/F program office, the most important source of savings on the main airframe MYP contract with Boeing was the $200 million in EOQ and CRI funding and annual AP funding. Significant price reductions from suppliers were achieved from EOQ funding. According to NAVAIR, however, CRI funding ($115 million) was a much more important source of cost savings than EOQ funding, and it provided a higher return on investment.

F/A-18E/F MYP II

In December 2003, the U.S. Navy awarded a second five-year MYP contract to Boeing valued at $8.9 billion. The contract originally envisioned procurement of 222 aircraft (later reduced to 210), made up of 154 F/A-18E/F aircraft and 56 EA-18G electronic attack aircraft (which

---

shown procured as follows: four in FY 2003, and four each year from FY 2005 to FY 2008, for a total of 20 aircraft.

has an airframe identical to F/A-18E/F). The contract spans the period of FYs 2005 through 2009. According to the February 2003 justification, the proposed MYP contract includes a cost savings of $1.052 billion, or an estimated 10.95 percent savings over single-year procurements estimated to total $9.612 billion. According to NAVAIR, the savings for MYP II represent in essence a 10.95 percent drop in average unit price from the last MYP I lot, with the price essentially remaining unchanged from that point on (in constant FY 2000 dollars).

Unlike MYP I, which was an FPI contract with a 70:30 split share, the MYP II contract is an FFP contract using a price-based acquisition (PBA) strategy, with TINA waivers and with minimal cost reporting and cost/pricing insight. The intention was a full transfer of the responsibility for realizing CRIs and the target price to the contractor. The incentive for the contractor to lower costs was that any under-runs would be 100 percent retained by the contractor.

According to the justification and NAVAIR, all contract cost savings were expected to come from CRIs, most of which were identified during the ongoing "Must Cost" initiative undertaken during MYP I. The contract provided $100 million investment funding for CRIs but no EOQ funding whatsoever.

The next subsection provides some overall observations on our review of the historical examples since 1982, with a strong emphasis on the seven post-1995 programs.

Observations on the High-Level Historical Overview of MYP Savings

Our high-level review of historical fixed-wing military MYP aircraft programs shows a wide variation in their characteristics, as shown in Tables 5.1 and 5.2. The vast majority (11 out of 14) of the contract savings percentage estimates in both periods, however, are rather significant and fall between 7 and 12 percent. For the early period, only the F-16 MYP III, with an estimated contract savings percentage of 5.7 percent, and the KC-10 at 17.7 percent, fall outside of this range. For the more recent programs, only the C-17 MYP I, at 5.5 percent, falls outside of this range.

Figure 5.1 summarizes the estimated contract savings percentages for all 14 programs.

Turning to the post-1995 programs, we notice that the key characteristics as laid out in Table 5.2 vary dramatically. For example, total procurement numbers vary from 8 to 210, while annual procurement rates vary from 2 to 42. MYP program lengths differ significantly, from four to seven years. EOQ funding covers a wide range, from 0 to 35 percent of estimated contract value. We also note that CRI-funding, at least as an integral part of MYP contracts, is rare. What limited data we have on the pre-1995 programs confirm this diversity of characteristics. For example, for the earlier programs, total procure-

Figure 5.1
Estimated MYP Contract Savings Percentage Compared with SYP Contracts for 14 Fixed-Wing MYP Programs, 1982–2005 (based on TY$ program estimates)

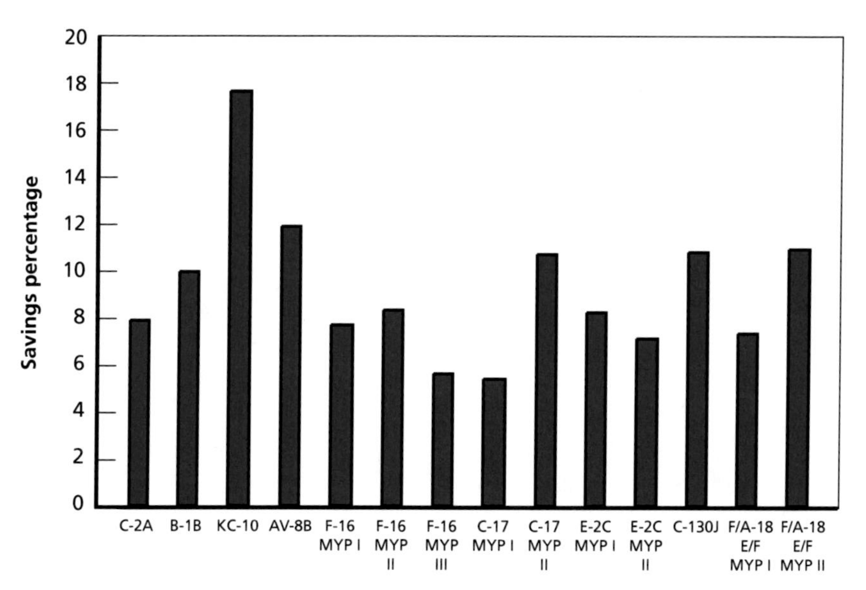

RAND *MG664-5.1*

Table 5.2
Post-1995 Fixed-Wing Military Aircraft MYP Contract Characteristics and Cost Savings Estimates Based on Initial MYP Justification Package

MYP Program	Airframe Contract Value (TY $billions)	Estimated Contract Savings (TY $billions & %)	Quantity: Total, Annual Rates	Time Frame (Total & FYs)	EOQ Total (TY $millions & % MYP contract)	CRI (Total in TY $ millions, % MYP contract)	MYP Production Rate Change vs. PYs	Program Maturity	Contract Type
C-17 I	$14.2	$0.9 5.5%	80 8–15	7 yrs. 97–03	$300 2%	$300–500, 2–3%[a]	Increase >85%	Start of FRP	FPIS + FFP
C-17 II	$9.7	$1.2 10.8%	60 15	5 yrs. 03–07	$645 7%	$200–300, 2–3%[b]	No change	Mature	FFP-EPA PBA-TINA waivers
E-2C I	$1.3	$0.11 8.3%	21 3–5	5 yrs. 99–03	$418 35%	None	Slight increase	Mature (air-frame)	FFP
E-2C II	$0.8[c]	$0.06[c] 7.2%	8 2	4 yrs. 04–07	$85.8 10.9%	None	Large decrease	Mature	FFP
C-130J	$4.0	$0.5 10.9%	60 4–13	6 yrs. 0308	$140 3.5%	None (?)	Varies	Early FRP	FFP + EPA FAR Parts 12 to 15

Table 5.2—continued

MYP Program	Airframe Contract Value (TY $billions)	Estimated Contract Savings (TY $billions & %)	Quantity: Total, Annual Rates	Time Frame (Total & FYs)	EOQ Total (TY $millions & % MYP contract)	CRI (Total in TY $millions, % MYP contract)	MYP Production Rate Change vs. PYs	Program Maturity	Contract Type
F/A-18 E/F I	$8.9	$0.7 7.4%	210 36–42	5 yrs. 00–04	$85 1%	$115, 1%	Increase	Start of FRP	FPI 70:30 split
F/A-18 E/F II	$8.9	$1.1 10.95%	210 42	5 yrs. 05–09	None	$100, 1%	No change	Mature, major ECPs	FFP, PBA-TINA waivers

NOTES: FPIS = Fixed Price Incentive, Successive Targets; FFP = Firm Fixed Price; EPA = Economic Price Adjustment; TINA = Truth in Negotiation Act; PBA = Price Based Acquisition.

[a] CRI was funded under totally separate contracts that often overlapped but did not exactly coincide with the periods covered by the two C-17 MYP contracts.

[b] Some CRI funding was under separate contract with varying time frames.

[c] Includes engine and airframe.

ment numbers varied from 39 to 780, and EOQ funding varied from 2 percent to 17 percent of estimated contract value.

We conducted some elementary correlation analysis of the relationship between the estimated contract savings percentage and the program characteristics listed in Table 5.2, variations in which, we suggested at the beginning of this chapter, might be expected to affect the scale of estimated savings. However, we found no statistically significant correlations between savings percentage estimates and program characteristics, with the partial exception of one factor. If the C-17 MYP I program is excluded, we found a relatively weak but probable correlation between estimated contract savings percentage and the duration of the MYP program. However, if the savings estimates are changed to constant FY 2005 dollars and the savings percentages recalculated (as shown later in Table 5.4), even this weak correlation disappeared.

In summary, our analysis failed to identify any significant correlations whatsoever between the magnitude of the savings percentage estimate and the various program characteristics listed in Table 5.2, such as total procurement numbers, annual production numbers, size of contract, absolute and relative amounts of EOQ funding, stage of production maturity, and so forth. Thus, these characteristics are apparently not useful in predicting the relative scale of savings percentage estimates. A very mature program with very small procurement numbers, no formal CRI funding, and set up with a short MYP program duration, such as the E-2C MYP II, may result in a savings percentage estimate roughly the same as a relatively less mature, longer MYP program with high procurement numbers and significant CRI funding, such as the F/A-18E/F MYP I.

However, it is important to point out that this does not necessarily prove that no correlations exist among these or other factors and the magnitude of estimated savings. The problem is that too few data points exist to support a statistically significant correlation analysis. Correlations may exist, but they cannot be statistically demonstrated with this limited database.

From our more detailed analysis of the seven more recent MYP programs, including discussions with acquisition officials involved with many of the programs, we were able to glean some additional observations. In most cases, program documents and officials identified supplier and vendor discounts as the most significant source of cost savings, followed closely by process improvements. Virtually every program identified EOQ and CRI funding as crucial enablers for cost reductions. However, the estimated relative importance of EOQ compared with CRI funding, as well as the scale of such funding, varied widely from program to program.

CRI funding was often, but not always, emphasized as the most important enabler for cost savings. The program offices considered CRI funding to be crucial for the cost savings on the F/A-18E/F I and II MYP programs, as well as on the C-17 I and II MYP programs. However, the linkage between MYP and CRI funding is unclear. CRI funding can be, and often is, provided on programs without MYP contracts. The CRI funding that was so important in reducing costs on the C-17 program was provided by contracts separate from the MYP contracts and was used extensively before the MYP contracts. In addition, no formal CRI funding was provided at all on the E-2C I and II MYP programs, or on the C-130J MYP program. Thus, it is difficult to make any broad generalizations about the linkage and importance of CRI funding for the magnitude of MYP program savings percentage estimates.

Program officials whom we interviewed had mixed views about the importance of EOQ funding. Based on the data we collected, EOQ funding is more common than CRI funding on MYP contracts and often accounts for a larger percentage of the MYP contract compared with CRI funding. NAVAIR officials regarded EOQ funding on both the E-2C I and II MYP programs as absolutely crucial for obtaining significant cost savings. However, officials on the F/A-18E/F MYP I program insisted that EOQ funding yielded disappointing results, leading to no request for EOQ funding for the second F/A-18E/F MYP program. In addition, we were told that some EOQ funding on the F/A-18E/F MYP I program, as well as on the C-130J program, was redirected into CRI efforts at the subcontractor level.

Our interviews suggested that prime contractor leverage with and management of vendors and suppliers were crucial, regardless of the amount of EOQ funding made available. That leverage arose in part from the broad business environment that existed when the programs were undertaken, as well as the motivation and incentives of the prime contractor to work with vendors to ensure the best possible outcome for the customer.

What do the historical case studies tell us, if anything, about the reasonableness of the estimated F-22A MYP savings? While the estimated savings for the F-22 MYP as a percentage of the total contract appear relatively small in historical terms, we can also see that in many key aspects the F-22A MYP differs significantly from historical cases. In Figure 5.2, we present a summary of our qualitative assessment of various key aspects of historical MYPs compared with the proposed F-22A MYP.[16] We identify seven key characteristics that in theory should affect the relative scale of savings that can be expected from a MYP program. We then show a qualitative comparison with the F-22A proposed MYP of each of the seven factors for each of the seven recent historical MYP programs. These comparisons are color-coded. Green means the program characteristic for the F-22A program was, at least in theory, more conducive to producing relatively higher savings as a percentage of the contract than historical programs. Blue means that the conditions for the F-22A program were roughly equivalent to the historical programs. Finally, red means the conditions for the F-22A program were in theory less conducive to savings compared with historical programs.

As demonstrated by the "stop light" chart, most factors for the recent historical fixed-wing MYPs were more conducive to savings compared with the F-22A. For some factors, such as EOQ timing (the point at which EOQ funding is made available) and time frame, *all* the historical programs had more favorable conditions than the F-22A proposed MYP. Only the E-2C II MYP program had more than one factor that was less favorable to savings than the F-22A program, and it

[16] The key characteristics of the MYP program that contribute to cost savings are explained in the box on pages 50–51, except for change in production rate, which is an indication of the scale and direction of the anticipated change in annual production rate when the annual rate planned for the MYP program is compared with the annual production rate of the fiscal year immediately prior to the beginning of the MYP program.

Figure 5.2
Comparison of the Seven Recent Historical MYP Programs with the Proposed F-22A MYP

MYP program	Total number/ rate	EOQ funding	EOQ timing	Time frame	CRI funding	Program maturity	Change in production rate	Contract savings
C-17 I								$0.9 5.5%
C-17 II								$1.2 10.8%
E-2C I								$0.1 8.3%
E-2C II								$0.1 7.2%
C-130J								$0.5 10.9%
F/A-18 E/F I								$0.7 7.4%
F/A-18 E/F II								$1.1 10.95%

More favorable for F-22 savings compared to historical programs
Equally favorable for F-22 savings compared to historical programs
Less favorable for F-22 savings compared to historical programs

NOTE: Contract savings have been rounded to one decimal place.
RAND *MG664-5.2*

is considered to be a very unique case. Also note that both F/A-18E/F programs have almost all favorable characteristics compared with the F-22A MYP, with the exception of EOQ funding, which was minimal on F/A-18E/F MYP I and nonexistent on F/A-18E/F II.

While we recognize that this comparison is very general, high-level, and subjective, we nonetheless believe it makes a valid point. In many respects, the proposed F-22A MYP program differs significantly from historical programs in ways that make large savings percentages from an MYP less likely, at least in theory.

Comparisons of Savings Estimates per Aircraft

To round out our high-level assessment of historical multiyear programs, we adjusted all of the MYP contract estimated savings to FY

2005 dollars and calculated how many FY 2005 dollars were saved per aircraft on each program. In the first instance, this was done to help us conduct a more credible correlation analysis of the estimated MYP program contract savings percentages compared with other program characteristics, as discussed above. In addition, some observers have attempted to answer concerns expressed by those in Congress and elsewhere that the expected contract savings percentage on the proposed F-22A MYP program is small by historical standards, by noting that the dollar savings per aircraft on the F-22A MYP program are consistent with past MYP programs. To examine this hypothesis more closely, we recalculated contract savings estimates and dollar savings per aircraft in FY 2005 dollars for our 14 historical cases. The results of this effort for the early programs from 1982 through 1995 are shown in Table 5.3; those for the post-1995 programs are shown in Table 5.4.[17]

As Table 5.3 shows, savings per aircraft in 2005 dollars for the early period varied between $680,000 for the F-16 MYP II and $22 million for the KC-10. Removing the KC-10 and the B-1B shows that the remaining five programs saved roughly $1–2 million per aircraft. The F-16 MYP programs saved between about two-thirds million and one million 2005 dollars per aircraft. This compares favorably with estimates for the proposed F-22A MYP program. However, with a much smaller unit procurement price, the F-16 savings represent a much higher percentage of SYP unit costs than do the estimated F-22A MYP per aircraft savings.

[17] All conversions to FY 2005$ were done using the Air Force Aircraft Procurement Index for President's Budget Year 2007, rebaselined to FY 2005 dollars. Annual savings data were available for the F-16 MYP II; B-1B, C-2A, and C-17 MYP II; E-2C MYP I and II; C-130J, and F/A-18E/F MYP I and II programs. Converting to FY 2005 dollars was done by deflating the savings in each year. We had only a total savings estimate for the F-16 MYP I and III, and the KC-10, AV-8B, and C-17 MYP I programs. For these, we calculated a weighted deflator using the annual procurement funding from the most representative SAR as the weights. The F-16 MYP I used the 12/82 SAR, the F-16 MYP III used the 12/88 SAR, the KC-10 used the 6/83 SAR, the AV-8B used the 12/88 SAR, and the C-17 MYP I used the 12/05 SAR. The annual quantities in the SARs match the quantities in the MYP savings estimates in all cases except for the last year of the C-17 MYP I; there, we made a simple ratio adjustment.

Table 5.3
Contract Savings, in Millions of TY$ and 2005$, and Savings per Aircraft, for MYP Programs, 1982–1995

MYP Programs 1982–1995	Total TY$ Estimated Savings	Total 2005$ Estimated Savings	2005$ Estimated Savings per Aircraft
F-16 I	246.0	416.4	0.87
F-16 II	358.3	486.2	0.68
F-16 III	421.9	636.6	1.06
KC-10	600.0	967.9	22.00
B-1B	1,188.2	1,754.9	19.07
C-2A	58.4	79.1	1.80
AV-8B	124.0	165.3	2.30

Table 5.4
Contract Savings, in Millions of TY$ and 2005$, and Savings per Aircraft, for MYP Programs, 1995–2005

MYP Programs 1995–2005	Total TY$ Estimated Savings	Total 2005$ Estimated Savings	2005$ Estimated Savings per Aircraft
C-17 I	900.0	974.1	12.18
C-17 II	1,211.0	1,164.8	19.41
E-2C I	106.5	102.7	4.89
E-2C II	61.3	63.0	7.87
C-130J	340.1	314.9	7.50
F/A-18E/F I	706.1	748.7	3.37
F/A-18E/F II	1,052.3	1,002.2	4.77

Table 5.4 shows the savings comparisons for the more recent MYP programs. Here, the savings vary between $3.37 million per aircraft for the F/A-18E/F MYP I and nearly $20 million per aircraft for the C-17 MYP II. If the C-17 program is excluded, the savings range per aircraft narrows to from $3.37 million for the F/A-18E/F to just under $7.9 million for the E-2C II MYP program. The savings for the one fighter program, the F/A-18E/F, remain in the $3–4 million range. Again, the proposed F-22A savings estimates per aircraft do not compare unfavorably to these estimates. However, the unit price of the F-22A is of course higher than the F/A-18E/F.

Results of a Quantitative Analysis of the F/A-18E/F MYP I Program

We were able to obtain more detailed and complete cost data on the F/A-18E/F MYP I than for any other historical MYP program. We used these data both to attempt to verify whether the originally estimated savings were in fact achieved and for a more detailed comparison with the proposed F-22A program.

For the F/A-18E/F program we were able to obtain four types of data: the program Selected Acquisition Reports (SARs); the P-5 Cost Analysis Exhibits in the Justification of Estimates for the Department of the Navy Fiscal Year Budget Estimates; the Naval Air System Command's Historical Aircraft Procurement Cost Archive (HAPCA); and a few Contractor Cost Data Reporting (CCDR) system documents, specifically Cost Data Summary Reports (CDSRs, DD Form 1921) and Functional Cost Hours Reports (FCHRs, DD Form 1921-1). Thus, we believe we obtained sufficient data to permit a reasonably credible rough estimate of the likely savings on the F/A-18E/F program and to compare that estimate with the savings estimate in the original MYP program justification. The strengths and shortcomings of each type of data, as well as the methodology we adopted to evaluate the data, are discussed in detail in Appendix A.

Many factors hamper an after-the-fact assessment of whether a multiyear procurement approach actually saved costs, including availability of data, consistency between the content of the data and the content of the program subject to multiyear contracting benefits, and changes in the program after the decision to employ multiyear contracting and the completion of the contract. Keeping this caveat in mind, the analyses we were able to perform using the available data for the F/A-18E/F MYP I support the conclusion that savings were realized and the magnitude was probably in the neighborhood of the original justification estimate. This conclusion would likely not be changed if some of the missing data were available. A more definitive answer would require a much more detailed analysis at the individual cost reduction initiative and economic order quantity level.

Comparing the Proposed F-22A Multiyear Procurement and the F/A-18E/F Program

Some observers have expressed an interest in the possibility of normalizing existing F/A-18E/F MYP I data to make them more comparable to the quantity, rate, and duration assumptions in the proposed F-22A MYP program, thus facilitating more credible comparisons of savings percentages. We undertook such a normalization while fully recognizing that the widely differing circumstances surrounding both programs make such an effort extremely challenging. We attempted a relatively straightforward approach, fully understanding that the outcome would produce only very approximate high-level results that must be viewed with extreme caution. Nonetheless, we believe such an effort is instructive and useful for gaining additional insights into both programs.

The proposed F-22A multiyear procurement covers 60 aircraft over a three-year period. In terms of production aircraft sequence, these aircraft are equivalent to numbers 116 through 175 for the F/A-18E/F. We set out to estimate how much savings were associated with aircraft numbers 116 through 175 for the F/A-18E/F program. We calculated an optimistic estimate of this savings by constructing cost improvement curves for the original F/A-18E/F single-year and multiyear procurement costs and then regrouping aircraft 116 through 175 into three 20-aircraft buys.[18]

Table 5.5 shows the quantities and single-year and multiyear funding estimates for the F/A-18E/F MYP I time period. The table also shows the annual buy midpoints used for the cost improvement curve analysis. The midpoints account for the 62 aircraft produced in the three LRIP buys.

The then-year dollar values in Table 5.5 were then converted to FY 2005 constant dollars using Aircraft Procurement, Navy (APN) deflators. Unit cost improvement curves were determined for the

---

[18] The estimate is optimistic because at the time of the F/A-18E/F MYP I, the planned program had a total of 548 production aircraft, with 264 following the proposed multiyear contracting period. The F-22 program has no aircraft planned beyond the proposed multiyear contract.

Table 5.5
Annual Procurement Funding and Savings During the First F/A-18E/F Multiyear Contract (MYP I)[a]

Fiscal Years	Annual Quantities	Lot Midpoints	Single-Year Procurement	Multiyear Procurement	Savings
2000	36	80	3,071.297	2,923.772	
2001	42	119	3,183.322	3,020.125	
2002	48	164	3,365.072	3,199.452	
2003	48	212	3,280.519	3,156.650	
2004	48	260	3,402.869	3,296.975	
Totals	222	—	16,303.079	15,596.974	706.105 (4.3%)

[a] Based on February 1999 multiyear justification. All dollar values are in TY $millions.

single-year and multiyear funding estimates. The resulting curves were used to calculate the procurement funding required for aircraft 116 through 175. The funding values were adjusted to reflect the differences in production rates between the program shown in Table 5.5 and three annual buys of 20 aircraft.[19]

The results are shown in Table 5.6. The savings are TY $229.8 million compared with $706.1 million for the 222 aircraft shown in Table 5.5. In constant FY 2005 dollars, the 60 aircraft program savings amount to $233.8 million.

Notice that while the total dollar savings decline dramatically (from TY $706.1 million to TY $229.8 million), the estimated savings percentage remains about the same. As shown in Table 5.6, estimated savings on the actual F/A-18E/F MYP I program equaled about 4.3 percent of the estimated annual procurement funding for single-year procurement during the same period as the MYP I program.[20] The savings shown in Table 5.6, estimated for a hypothetical three-year

[19] We used a 90 percent rate slope adjustment, resulting in an increase of 12 percent for aircraft in the 42-per-year buy and 14 percent for aircraft in the 48-per-year buys.

[20] The official justification savings estimate of 7.4 percent for the F/A-18E/F MYP I was derived by comparing the total estimated savings (TY $706.1 million) to the estimated SYP contract value, which was put at TY $9,546.9 million for 222 aircraft.

Table 5.6
Estimate of Annual Procurement Funding for 60 F/A-18E/F Aircraft at 20 per Year Beginning with Aircraft Number 116 (TY $millions)

Year	Annual Quantities	Single-Year Procurement	Multiyear Procurement	Savings
2002	20	1,722.8	1,642.6	
2003	20	1,724.3	1,647.9	
2004	20	1,724.4	1,651.2	
Totals	60	5,171.5	4,941.7	229.8 (4.4%)

60-aircraft MYP program for the F/A-18E/F, represent an essentially comparable savings percentage of 4.4 percent.

Summary Conclusions

Our high-level survey of the seven major fixed-wing combat aircraft MYP programs from 1982 through 1995, our more detailed review of the seven additional MYP programs from 1995 through 2007, and our quantitative analysis of the F/A-18E/F MYP I data available to us showed the following:

- Historical MYP fixed-wing aircraft program contract savings percentage estimates from 1982 through 2007 varied from 5.5 percent to 17.7 percent. Estimated savings for fighter aircraft MYP programs during this period varied from 5.7 percent to 11.9 percent (all based on original MYP program justification estimates using TY$).
- For MYP programs after 1995, our quantitative analysis identified little or no significant correlation between the magnitude of the contract savings percentage estimates and any of the following factors: contract size, total procurement numbers, annual procurement numbers, program duration, EOQ, and CRI funding.

- Qualitative examination of the post-1995 MYP programs and many of the pre-1995 programs suggested that achieving subcontractor and vendor quantity discounts is a key factor in obtaining savings on MYP programs. Achievement of significant savings in this area, however, did not always appear to depend on EOQ funding. Often, the general business and political environments and the motivation and incentives for the prime contractor and subcontractors to achieve savings were more important. Many programs, although not all, identified process improvements and other CRIs, whether government funded or not, as key contributors to significant cost savings.
- Our qualitative assessment and our comparison of key program MYP attributes of the seven most-recent historical aircraft MYPs—attributes that in theory should contribute to greater MYP savings—with those same attributes on the F-22 MYP, suggested that the F-22 MYP is at a relative disadvantage and cannot be expected to achieve contract savings as high as the earlier historical programs.
- Our quantitative analysis of the F/A-18E/F MYP I data suggested that the program likely has achieved contract savings that are generally consistent with the original program justification savings estimates.
- Our attempt to normalize the F/A-18E/F MYP I data to make them more comparable to the proposed F-22A MYP program, while admittedly rather imprecise and optimistic, indicated that a hypothetical three-year MYP might have produced far less total dollar savings than the actual program, but those savings probably would have represented roughly the same savings percentage as the baseline single-year procurement estimates.

CHAPTER SIX

Results and Findings

This chapter brings together the findings and conclusions of previous chapters and presents our results.

Estimating the three single-year contract prices and comparing them to the multiyear contracts for air vehicles and engines recently negotiated between the USAF and the F-22A contractors produce a range of estimates. We estimate the MYP savings to range between $274 million and $643 million, based on historical F-22A cost data and three alternative cost improvement curve assumptions. When using the second assumption for predicting these costs (the assumption that Lot 5 and 6 costs were representative of Lot 7 through 9 costs), we found that our prediction and the separate, negotiated single-year Lot 7 price were similar. This method resulted in overall savings of $411 million, or 4.5 percent of the multiyear contract.

To help provide perspective on the realism of our savings estimate, we also substantiated savings initiatives proposed by the F-22A contractors. We assessed these itemized, specific, and documented efforts, and we traced the savings to the final negotiated contract. This approach produced savings estimates of $296 million. Thus, over 70 percent of our estimated savings of $411 million can be traced to these proposed initiatives.

Finally, although the savings percentage (4.5 percent) for the F-22A multiyear contract compared with the single-year predicted prices was relatively low by historical estimates, the savings per aircraft in the F-22A multiyear contract were on the upper end of the estimated savings of previous fighter/attack multiyear contracts (in part due to the higher unit cost of the F-22A). Also, our qualitative assessment of historical program attributes that may contribute to greater percentage cost savings estimates showed that the F-22A MYP differs considerably from most historical programs and is in many respects unique.

Calculation of the Savings Percentage

To calculate the savings percentage, we divided the assessed savings by our estimate of the single-year contract for Lots 7, 8, and 9. These savings percentages are listed in Table 6.1.

We define savings percentage as

Total Amount of Savings / Total of Single-Year Estimates.

Table 6.1 displays the savings percentages if we use our estimate of savings from the difference between our single-year cost estimates and the multiyear contract negotiated values.

As can be seen, our substantiated savings ($296 million) for the F-22A firm-fixed-price multiyear contract is more than 70 percent of our estimated savings of $411 million.

Figures 6.1 and 6.2 show additional ways of portraying the savings of the multiyear procurement. Figure 6.1 arrays the results of our research as a function of the percentage of savings relative to a single-year procurement. The left bar shows historical data for all fixed-wing aircraft procured under a multiyear contract since 1982, indicating

Table 6.1
Savings Percentages Using RAND's Estimated Savings (SYP Minus MYP)

CIC Assumption	Single-Year Estimate (TY $millions)	Savings SYP Minus MYP (TY $millions)	Savings Percentage
Lots 1–6	8,952	274	3.1
Lots 5 and 6	9,089	411	4.5
Lot 6	9,320	643	6.9

Figure 6.1
Estimated Savings Percentage Relative to SYP Value

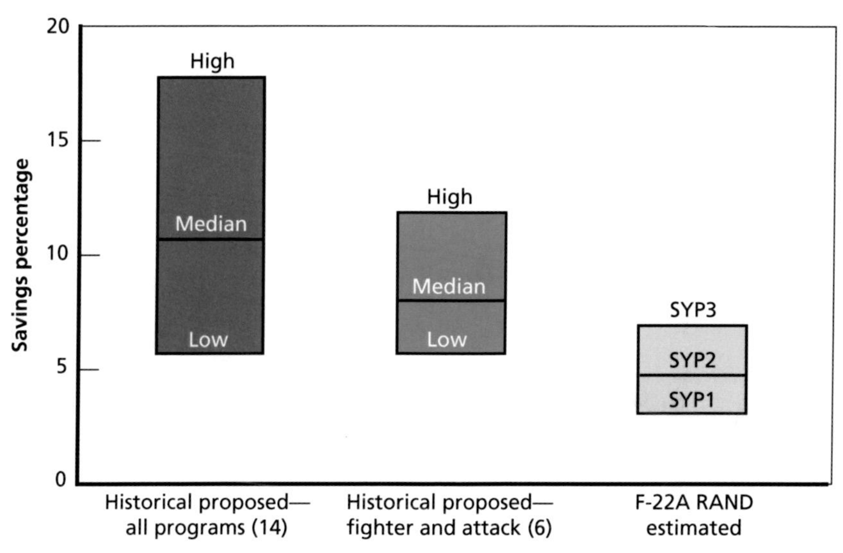

NOTE: The numbers in parentheses in the first two bar labels show the number of aircraft programs included in the analysis.

RAND *MG664-6.1*

the low, median, and high values. The next bar presents the data only for fighter and attack aircraft, displayed in a similar fashion. Finally, the right bar shows the savings calculated based on assumptions about SYP learning curves (SYP1 reflects our first assumption, SYP2 reflects our second assumption, and SYP3 reflects our third assumption). The chart shows that the substantiated savings fall within the range of our estimated savings but that they are low by historical estimates both for all aircraft programs and for only fighter aircraft.

A second way to evaluate savings is by the number of dollars saved per aircraft. Figure 6.2 contains a display similar to Figure 6.1, except that the metric is dollar savings per aircraft in FY 2005 dollars.

Figure 6.2
Estimated Dollar Savings per Aircraft

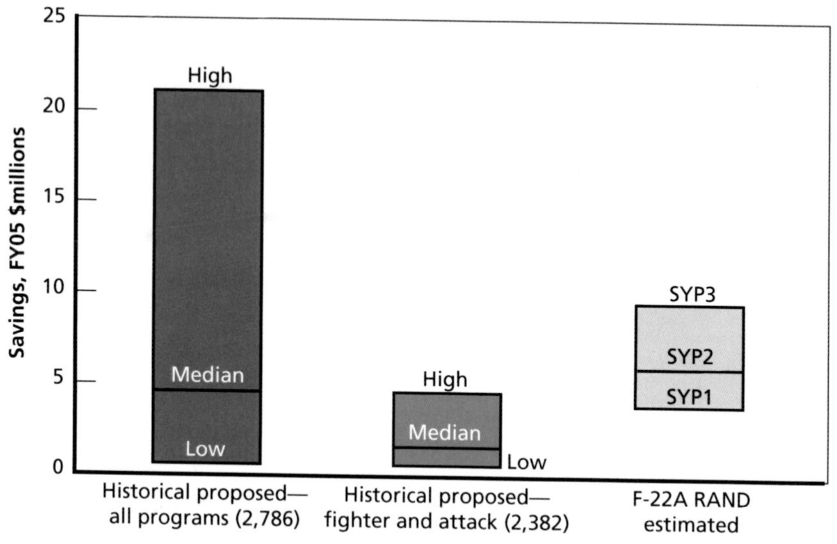

NOTE: The numbers included in the parentheses in the first two bar labels show the number of aircraft included in the analysis.

RAND *MG664-6.2*

In this case, the substantiated savings also fall within the range of our estimated savings but they are high by historical estimates for fighter aircraft, which is partially explained by the higher unit cost of the F-22A compared with other fighters. However, most of the factors that contribute to savings for the most recent historical fixed-wing MYPs were more conducive to savings than is the case for the F-22A. For some factors, such as EOQ timing (the point at which EOQ funding is made available), and time frame, *all* the historical programs had more favorable conditions than the F-22A proposed MYP.

APPENDIX A

After-the-Fact Analysis of F/A-18E/F Multiyear Savings

Frequently, after the multiyear contract has been implemented, the question arises as to whether there were actual savings. We sought to address this question using F/A-18E/F cost data for the various annual procurements over the course of the lead-in single-year procurements (SYP) and the following years of the MYPs.

Approach

The general approach is to fit a cost improvement curve to the lead-in lots. This curve is projected over the quantities of the following lots, and the projected total cost of the following quantities is calculated and compared with their "actual" costs. "Actual" is shown in quotation marks because frequently the costs of the following quantities are estimates ahead of their realization.

Available Data

For the F/A-18E/F program, we were able to obtain four types of data: the program Selected Acquisition Reports (SARs); the P-5 Cost Analysis Exhibits in the Justification of Estimates for the Department of the Navy Fiscal Year Budget Estimates; the Naval Air System Command's Historical Aircraft Procurement Cost Archive (HAPCA); and a few Contractor Cost Data Reporting (CCDR) system documents, specifically Cost Data Summary Reports (CDSR, DD Form 1921) and Functional Cost Hours Reports (FCHR, DD Form 1921-1). Because of the procedures used to prepare these documents and data collections, we did not receive a sufficient number of observations (fiscal years) nor all of the necessary insight into the elements of the program's work breakdown structure (WBS) to do a thorough analysis regarding the realization of "actual" savings.

The available data are summarized in Table A.1. The low rate initial production (LRIP) phase is shown in blue, the first multiyear procurement phase is shown in red, the second multiyear procurement phase is shown in green, and the following years are shown in orange.

Selected Acquisition Reports (SARs)

The SARs provide the annual flyaway funding profile in program base-year dollars and annual procurement funding profiles in program base-year and then-year dollars. Flyaway costs are broken out into nonrecurring and recurring costs, but no further detail is provided. It is

Table A.1
Available Data Sources by Fiscal Year and Program Phase for the F/A-18E/F MYP I and II Programs

Fiscal Year	Phase	SAR	P-5	HAPCA	CCDR-Boeing	CCDR-NGC
1997	LRIP	X	X	X		
1998	LRIP	X	X	X		
1999	LRIP	X	X	X	X	
2000	MYP I	X	X	X	X	X
2001	MYP I	X	X		X	X
2002	MYP I	X	X		X	X
2003	MYP I	X	X		X	X
2004	MYP I	X	X			
2005	MYP II	X	X			
2006	MYP II	X	X			
2007	MYP II	X	X			
2008	MYP II	X	X			
2009	MYP II	X	X			
2010		X				
2011		X				

important to keep in mind that the annual procurement funding is for the *total* program. It includes the costs of (1) the prime contractor (Boeing/McDonnell Douglas) and all their subcontractors (including Northrop-Grumman), (2) all companies that contract directly with the government (including General Electric) and provide what is termed *government furnished equipment* (GFE), and (3) all government costs that are counted directly against the program, such as the program office, test facilities, etc. We analyzed data from the F/A-18E/F SARs for December 1997 and December 2005.

Exhibit P-5 Cost Analysis

Budget documents for the next fiscal year (say, FY 2008) are generally submitted in February of the current year (say, February 2007). The Navy P-5 exhibits provided a detailed breakout of the requested annual funding for the current year (FY 2007), the prior year (FY 2006), the sum of all prior years (FY 2005 and earlier), the budget year (FY 2008), and—depending on the budget cycle—the following budget year (FY 2009).[1] The P-5 data are updated each fiscal year to reflect committee actions, etc., so it is important to "work backward" from the most current data to the beginning of the program.

The detailed breakout consists of 29 elements, some of which are subtotals and a few of which are not used (blanks). These elements generally match up with the 41 elements in the HAPCA database (see below). For the purposes of this study, we analyzed the Airframe/CFE element and the procurement element. Airframe/CFE relates closely to the airframe prime contractor efforts covered in the multiyear agreement, and the procurement cost matches up with the procurement cost from the SAR.

Historical Aircraft Procurement Cost Archive (HAPCA)

Naval Air Systems Command maintains an archive of historical aircraft procurement costs with 41 elements that sum to procurement cost. Several elements are subtotals and some are blank (reserved for

---

[1] The FY 2008 budget documents show both FYs 2008 and 2009. The FY 2007 budget shows only the FY 2007 values and the priors but not the following budget year.

future use). The costs are all in then-year dollars. We analyzed Airframe/CFE and procurement costs for FYs 1997–2000 (the most recent available).

Contractor Cost Data Reporting (CCDR) Documents

The Contractor Cost Data Reporting (CCDR) system accumulates actual contractor cost data for major weapon system development and procurement programs.[2] The CCDR system has three primary types of documents: (1) the "Cost and Software Data Reporting Plan" (CSDRP, DD Form 2794), (2) the "Cost Data Summary Report" (CDSR, DD Form 1921), and the "Functional-Cost Hour and Progress Curve Report" (DD Form 1921-1).[3]

For the F/A-18E/F program, we were able to obtain CDSRs for the Boeing (McDonnell Douglas) program for (1) FY 1999, as of September 30, 2001 (approximately 88 percent complete); (2) FY 2000, as of July 31, 2002 (approximately 98 percent complete); (3) FY 2001, as of July 31, 2003 (approximately 89 percent complete); (4) FY 2002, as of June 30, 2004 (approximately 95 percent complete); and (5) FY 2003, as of July 28, 2005 (approximately 88 percent complete).

The latter four documents also provided Northrop-Grumman's total cost.

Analysis Approaches

Development of savings from multiyear procurement instead of single-year procurement requires estimating the costs for both the single-year and multiyear contracting approaches. Prior to execution of a multiyear contract, all these values are necessarily estimates. After execution of the multiyear contract, only the actual results of that contract can be collected and then compared with estimates of the single-year

[2] The specific requirements for reporting are described in DoD Instruction 5004M-1. The most recent issue is a draft dated February 2004.

[3] For contracts signed before October 1, 2003, the "Functional Cost Hour Report" (FCHR, Form 1921-1) and the "Progress Curve Report" (Form 1921-2) were separate reports.

approach developed years earlier. That can be either the estimate in the original justification or an alternative estimate. When using data sources that do not align with procurement cost or contract cost, an alternative estimate must be used.

During the course of execution of a multiyear contract there may be changes to the program. Quantities and production rates may increase or decrease, model mixes may be changed, capabilities may be added or removed, etc. Comparisons of multiyear actual costs with the original single-year estimate may require adjustments. Calculating these adjustments may range from relatively straightforward to impossible without detailed data.

Most multiyear procurement justifications provide annual funding estimates for single-year and multiyear alternatives and the savings, for both the weapon system (procurement less initial spares) cost of the system and the contract value. The initial spares funding in the budget justification can be added to the weapon system funding to obtain total procurement funding, which can be compared with procurement funding in Selected Acquisition Reports (SARs). One analysis approach we examined was to compare SAR procurement funding with the multiyear justification procurement funding. The SAR funding profile prior to the decision to award the multiyear contract can be used as a basis for projecting single-year procurement requirements and then compared with the values shown in the SAR subsequent to the MYP decision, or the latter can be compared with the single-year estimate in the multiyear justification.

Analysis Results

The F/A-18E/F program has had two multiyear procurements: MYP I, covering FYs 2000–2004, and MYP II, covering FYs 2005–2009. MYP II is less that half completed and many data sources are not available; hence, the discussion here will focus on MYP I.

MYP I—SAR Analyses

Before describing the analyses and results, it is important to explain how the estimates in Exhibit MYP-2, Total Program Funding Plans, were adjusted to match SAR procurement values. Table A.2 shows the F/A-18E/F Annual Procurement funding as shown in the February 1999 multiyear justification.

The multiyear justification is based on the funding requirements for the multiyear contracting period, in the present case FYs 2000 through 2004. The advanced procurement (AP) funding for FY 2000 is funded in FY 1999. The multiyear justification shows zero current year (CY) AP in FY 2004 because those funds would be for FY 2005, which is beyond the multiyear contracting period. The CY AP in FY 2004 is indicated by italics because it is not part of the multiyear justification. The FY 2004 weapon system cost is also italicized because it includes the CY AP value. Prior year (PY) and CY AP funding cannot be determined from the procurement funding profile presented in the SARs. In addition, the SARs include initial spares funding in the total procurement, which also cannot be broken out. Hence, the single-year and multiyear estimates must be adjusted as shown in Table A.2 to be comparable to the procurement values in the SARs. The bottom row in Table A.2 matches the values in the December 1998 F/A-18E/F SAR.

Table A.2
February 1999 F/A-18E/F Exhibit MYP-2 Multiyear Procurement Funding and Modifications to Match SAR Procurement (TY $millions)

	FY 1999	FY 2000	FY 2001	FY 2002	FY 2003	FY 2004
Gross cost		2,801.110	2,894.914	3,097.939	3,129.677	3,263.000
Less PY AP		–109.119	–105.529	–113.583	–111.520	–109.258
Net procurement		2,691.991	2,789.385	2,984.356	3,018.157	3,153.742
Plus CY AP	109.119	162.240	101.208	89.352	87.090	*88.903*
Weapon system cost	109.119	2,854.231	2,890.593	3,073.708	3,105.247	*3,242.645*
Plus initial spares		69.543	129.531	125.744	51.403	54.330
Procurement		2,923.774	3,020.124	3,199.452	3,156.650	3,296.975

Two multiyear procurement justifications were prepared for the F/A-18E/F MYP I program. The first is dated September 1998 and is out of cycle with the President's Budget. The second is dated February 1999 and was included in the FY 2000 President's Budget documentation. Both of these justifications showed nearly the same amount of expected savings, as shown in Table A.3. The percentage values are calculated relative to the single-year estimate. The percentages presented in Exhibit MYP-1, Multiyear Procurement Criteria, are relative to the estimated single-year contract value (for the September 1998 justification, the value is 7.4 percent).

The December 1997 SAR was the last produced prior to the September 1998 justification. We used the procurement cost data for FYs 1997–1999 (low rate initial production, LRIP) to construct a projection of the cost for the next five years to compare with the annual values in the SAR, which already reflected the proposed multiyear procurement. The results are shown in Figure A.1. The total costs and savings are summarized in Table A.4. The results compare very favorably with the multiyear justifications.

Table A.3
F/A-18E/F Multiyear Justification Procurement Cost Estimates (TY $millions)

	September 1998 Multiyear Justification	February 1999 Multiyear Justification
Single-year estimate[a]	16,239.594	16,303.080
Multiyear estimate[a]	15,529.195	15,596.975
Savings estimate	710.399 (4.4%)	706.105 (4.3%)

[a] Estimate includes initial spares and advance procurement funding for FY 2004, and exclude AP funding for FY 1999, as described above.

Figure A.1
Single-Year Procurement Projection and Estimated Multiyear Unit Procurement Cost from December 1997 SAR (FY 2005 $millions)

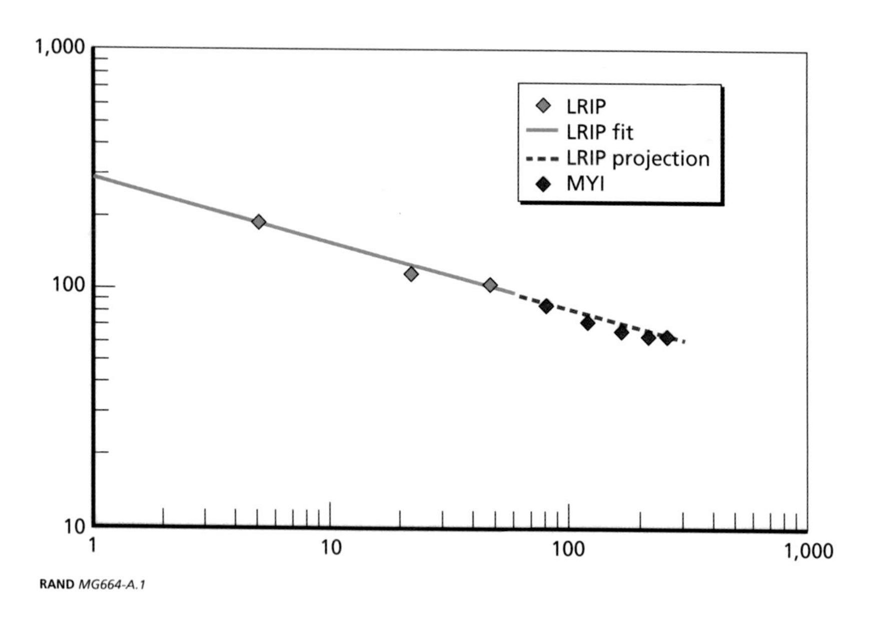

Table A.4
Estimated Single-Year Procurement Cost Compared with Sum of Multiyear Budget Values from December 1997 SAR (TY $millions)

	December 1997 SAR
Single-year projected cost	16,253.0
Multiyear budgeted value	15,550.2
Savings	702.8 (4.3%)

Were Estimated Savings Actually Achieved?

The question of interest is whether the program *actually* saved money, which requires analysis of the most recent SAR. Before proceeding with that analysis there are two important observations.[4] First, it is not meaningful to repeat the analysis shown in Figure A.1 and Table A.4 because the single-year projected cost obtained by analyzing the December 2005 SAR is not representative of the single-year program envisioned at the time of the multiyear justification. McDonnell Douglas began implementing cost savings initiatives prior to the start of the multiyear procurement and these steepened the slope of the fitted LRIP unit cost curve, resulting in a much lower projected cost during the multiyear contract.

Second, the cost during the multiyear contract period is not comparable to the original estimate, because the quantities of aircraft procured in three of the five years decreased, with a total reduction of 12 units, as shown in Table A.5. This change can be addressed by adjusting the quantity profiles used in the multiyear justification.

We converted the annual procurement values in the February 1999 MY justification from then-year to constant FY 2005 dollars for both the single-year and multiyear estimates. We fit a unit cost improvement curve to each set of numbers and regrouped them to match the actual annual quantities. We also applied a production rate adjustment. The resulting single-year, multiyear, and savings estimates for the actual 210-aircraft program are shown in Table A.6.

Table A.5
Annual Procurement Quantities at Time of the MYP I Justification and Actuals

	FY 2000	FY 2001	FY 2002	FY 2003	FY 2004
MYP I justification	36	42	48	48	48
December 2005 SAR	36	39	48	45	42

[4] There are probably several observations that may influence the results, but the two discussed here are readily addressed.

Table A.6
MYP I Savings Estimate Adjusted for Quantity and Rate Changes (TY $millions)

Adjusted to 210 Aircraft	February 1999 Multiyear Justification
Single-year estimate	15,586.3
Multiyear estimate	14,908.5
Savings estimate	677.9
	(4.3%)

The December 2005 SAR shows a total then-year dollar cost during the MYP I period of $15,578.5 million. This yields only a TY $7.8 million savings compared with the original single-year estimate adjusted to the actual quantities. Changes other than quantity occurred during the course of the program, and no doubt some of them were in cost elements that make up total procurement that were not part of the multiyear contract. Analysis of these changes would require far more detailed data than were available for this study. Consequently, we turn to other data sources that more closely match the multiyear contract content.

MYP I—P-5 Analyses

Data from the P-5 exhibits in the annual budget justifications permit selection of cost elements that may be close to the content of the multiyear contract. We analyzed Airframe/CFE (element 1), the sum of Airframe/CFE and CFE Electronics (element 5), and the sum of those two plus engineering change orders (ECO–element 12) and Nonrecurring Costs (element 14). The first two items are likely to be less than the contents of the multiyear contract and the third item is possibly more.

Because of uncertainties regarding the degree of alignment between these measures and the content of the multiyear contract, we cannot compare their values with the original single-year contract estimate. We performed the same type of analysis as shown in Figure A.1 and Table A.4 for the three P-5 measures. The cost improvement curve analysis for Airframe/CFE is shown in Figure A.2, and the funding and savings results are summarized in Table A.7. The results suggest that the savings achieved were of the same order of magnitude as originally estimated.

Figure A.2
Single-Year Procurement Projection and Estimated Multiyear Unit Airframe/CFE Cost from P-5 Exhibits

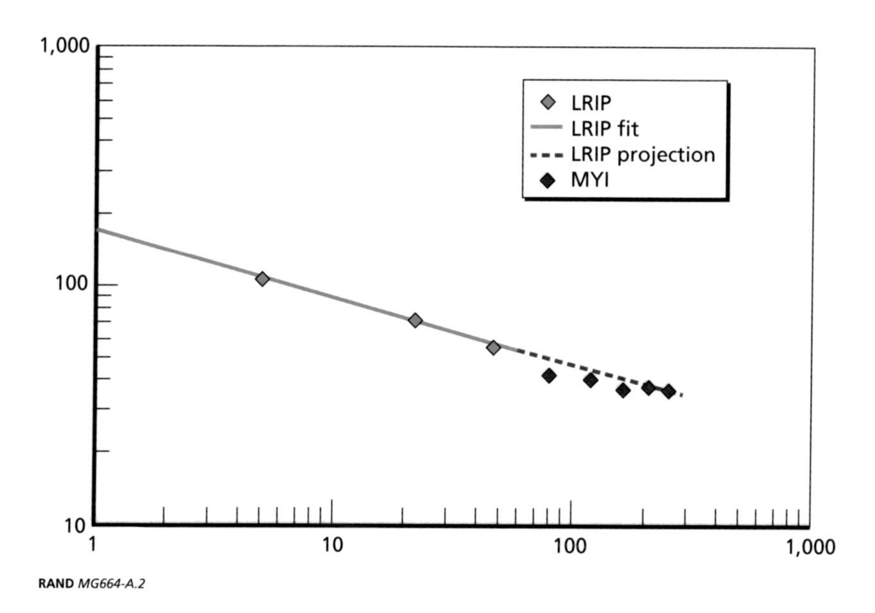

RAND *MG664-A.2*

Table A.7
Estimated Single-Year Procurement Cost Compared with Sum of Multiyear Budget Values from P-5 Exhibits (TY $millions)

	Airframe/CFE	Plus CFE Electronics	Plus ECO & Nonrecurring
Single-year projected cost	8,300.7	9,502.9	10,677.4
Multiyear budgeted value	7,833.5	8,947.2	9,559.7
Savings	467.2	555.7	1,117.7
	(5.6%)	(5.8%)	(10.5%)

MYP I—Other Analyses

We also considered the available HAPCA and CCDR data, but, as indicated by Table A.1, the coverage of LRIP and MYP I was incomplete and required several assumptions for us to proceed.

The HAPCA data cover the three LRIP years but only the first year of MYP I. This requires making assumptions about the final four of five years of MYP I. Additionally, comparing the HAPCA data with the P-5 data indicates that the HAPCA Airframe/CFE includes costs for many elements that are not included in the P-5 Airframe/CFE.

Boeing CCDR data were available for the last year of LRIP and the first four of the five years of MYP I. Using the last LRIP year as a reference point, making assumptions about the slope of the LRIP cost improvement curve, and projecting the final MYP I year based on the first four years, we found these data to be consistent with savings in the neighborhood of the original estimate. This conclusion assumes further that Northrop also achieved its share of savings.[5]

Summary Observations

After-the-fact assessment of whether a multiyear procurement approach actually saved money is hampered by many factors—including availability of data, consistency between the content of the data and the content of the program subject to multiyear contracting benefits, and changes in the program after the decision to employ multiyear contracting and the completion of the contract. The analyses we could perform using the available data for the F/A-18E/F MYP I support the conclusion that savings were realized and the magnitude was probably in the neighborhood of the original estimate. This conclusion would likely not change if some of the missing data were available. A more definitive answer would require a much more detailed analysis of the individual cost reduction initiatives and economic order quantity funding.

---

[5] With no Northrop data for the LRIP period, no comparative analysis of LRIP projections versus MYP I CDSR values could be done.

APPENDIX B

Reasons for Multiyear Savings from Previous Reports

This appendix contains a summary in tabular form of the reasons given for savings generated from multiyear contracts. The information was drawn from GAO reports, RAND research, and papers published by IDA. The table identifies the source, lists the sources of the savings, and provides the systems involved and the year of the multiyear purchase.

Source	Reasons Listed for Multiyear Savings and Other Issues Mentioned	System	Multiyear Buy (FY)
GAO/NSIAD-86-38 (February 1986)	More economical procurement of subsystems using EOQ funding than under annual procurements; broadening of the defense industrial base; more incentive to invest capital in new technology and modern equipment because of the long-term nature of a multiyear contract and industry desire to reduce manufacturing costs; protection against materials and parts lead-time increases. Contractors become more competitive in international sales, offer additional surge production output potential; lower procurement costs; increased capital investments; stable production schedules; retention of qualified, experienced staff to provide on the job training; increased surge capability; reduced administrative burden.	F-16	1982

Source	Reasons Listed for Multiyear Savings and Other Issues Mentioned	System	Multiyear Buy (FY)
GAO/NSIAD-88-233BR (September 1988)	Vendor and subcontractor items more economical under multiyear than under successive single-year contracts; majority of savings associated with procurement of vendor and subcontracted items on a more economical basis than possible with a series of annual procurements; multiyear contracting allows economic order quantity procurement; rather than procuring subcontracted parts and materials in annual lots of limited sizes, the prime contractor can procure parts in larger lots, thereby obtaining lower prices from subcontractors.	7 weapon systems	1990
GAO/NSIAD-86-5 (October 1985)	Advance materials cost growth avoidance; savings in contractor and government administrative costs; production efficiencies; program stability; increased investments in equipment or producibility; retention of skilled employees; better training programs.	MLRS	1982
GAO/NSIAD-86-1 (November 1985)	Multiyear business certainty conducive to the enhancement of the industrial base; improved competition in the subcontractor base increased quality and reliability and reduced prices; enhanced investment in facilities; improved vendor skill levels; initiation of new training programs; increased production capacity; more economical procurement of vendor and subcontracted items; efficiency in buying materials and scheduling production.	10 weapon systems	1986
IDA Paper-4116 (May 2006)	Procuring components at economic order quantities; reduced production line setups; reduced administrative procurement burden. (p. 18)	F/A-18	2002 (engines)

Source	Reasons Listed for Multiyear Savings and Other Issues Mentioned	System	Multiyear Buy (FY)
IDA Paper-4116 (May 2006)	Investment in longer-term capital equipment and manufacturing processes; reduced number of production line setups; reduced design engineering and configuration management recurring hours due to stable multiyear configuration; reduced administrative burden. (p.18)	F/A-18	2005
IDA Paper-4116 (May 2006)	Funded cost reduction initiatives; EOQ savings; supplier savings due to multiyear; new supplier for winglet and doors. (p. 19)	C-17	1997
IDA Paper-4116 (May 2006)	Efficiencies in planning and manufacturing resulting from stabilized production rate, longer term business arrangements with suppliers; reduced configuration variability. (p. 21)	C-130J	2003
GAO-89-224BR (September 1989)	The majority of the savings for DoD's multiyear contract candidates has been associated with procurement of vendor and subcontracted items on a more economical basis than is possible with a series of annual procurements. Multiyear contracting allows economic order quantity procurement. Rather than procuring subcontracted parts and materials in annual lots of limited sizes, the prime contractor can procure parts in larger lots, thereby obtaining lower prices from subcontractors.	8 weapon systems	1990
GAO-90-270BR (August 1990)	The majority of the savings for DoD's multiyear contract candidates has been associated with procurement of vendor and subcontracted items on a more economical basis than is possible with a series of annual procurements. Multiyear contracting allows economic order quantity procurement. Rather than procuring subcontracted parts and materials in annual lots of limited sizes, the prime contractor can procure parts in larger lots, thereby obtaining lower prices from subcontractors. Another significant source of savings is	6 weapon systems	1991

Source	Reasons Listed for Multiyear Savings and Other Issues Mentioned	System	Multiyear Buy (FY)
	attributed to manufacturing savings at the prime and major subcontractor levels. These savings result from such factors as improved fabrication, assembly, inspection, and test processes; reduced labor hours and spare part and repair requirements; and improved quality and reliability of the product. Inflation also saved 11.5%.		
GAO Testimony 114658 (March 10, 1981)	The contractor who holds a multiyear contract is able to spread his planning, startup, and other pre-production costs over a longer period of time, and more opportunity for increased efficiency and productivity should exist over this extended period. These contractor benefits should be transformed into decreased unit prices to the government. Likewise, administrative costs are saved by eliminating the costs attributable to repetitively soliciting and evaluating bids and awarding the contract. Still another advantage that has been repeatedly cited is that the quality of performance and service from contractors should increase. Contractor performance may be improved by reducing the uncertainty of continued government business; providing continuity in the delivery of recurring service and supply needs; and enabling the contractor to maintain a stable, well-trained workforce. Another advantage often cited by federal agency and contractor representatives is that multiyear contracting could lead to increased competition for government contracts. Many officials feel that with a longer time period for investment amortization allowed by the multiyear contract, a larger number of contractors, including small businesses, would be encouraged to compete for government contracts.	6 weapon systems	N/A

Source	Reasons Listed for Multiyear Savings and Other Issues Mentioned	System	Multiyear Buy (FY)
GAO-88-125 (May 1988)	Some prime contractors and many subcontractors cited multiyear contracting as a significant factor influencing their capital investment decisions. Two of the six prime contractors we reviewed told us that they would not have made any of their capital investments—totaling about $76 million—for these contracts had it not been for the advantages that multiyear contracts provided. About 81 percent of the 263 subcontractors reporting that they had made capital investments indicated that multiyear contracting has influenced these investments. The advantage of multiyear contracting most often cited was that it provides greater assurance of a stable, future defense business than annual contracting. Subcontractors are especially influenced by multiyear contracting because they generally view their annual contracts to be much less stable than their multiyear contracts. Subcontractors reported that prime contractors with multiyear contracts can often negotiate more effectively with their subcontractors because of the larger and more stable business base that multiyear contracts provide in comparison to annual contracts. Multiyear contracting has encouraged contractor investment, particularly at the subcontractor level, and has not limited price competition at the subcontractor level. Consequently, our findings suggest that multiyear contracting is a procurement technique that should contribute to improving the defense industrial base. (pp. 2–3)	N/A	N/A

Source	Reasons Listed for Multiyear Savings and Other Issues Mentioned	System	Multiyear Buy (FY)
	These contractors reported that multiyear contracting, unlike annual contracting, gives them greater assurance of the level of future defense sales, an important factor they consider when deciding how much to invest. As a result, many contractors reported that they purchased more and better capital equipment than they would have without multiyear contracts. (p. 17)		
RAND R-3990-DDR&E (Bodilly, Camm, and Pei, 1991)	The main benefit of multiyear contracting is that it changes the funding environment of contractors so that they are more likely to make decisions that will reduce procurement costs. The funding certainty generated by the production commitment enables contractors to use economic order quantities, reduced overhead, costs, and invest in new capital. A single large economic order quantity early in the program avoids the inflation cost of a series of annual orders of smaller quantities at inflated prices. (p. 7)	B-2	N/A
RAND N-1804-AF (Dews and Rich, 1982)	More dependable long-term production planning, productivity-increasing front-end investments in plant and training, more stable production rates, economies of scale in the purchase of materials and components, and (possibly) increased competition among suppliers. Larger contracts could stimulate additional firms to bid, as well as more aggressive bidding by those that do compete.	N/A	1982

Source	Reasons Listed for Multiyear Savings and Other Issues Mentioned	System	Multiyear Buy (FY)
GAO/T-NSIAD-96-137 (March 28, 1996)	Traditionally, a major portion of the savings from multiyear procurement comes from lower prices on economic order quantity buys from vendors or subcontractors. These savings are generally achieved with the EOQ money provided by the government. (p. 5)	C-17	1997

APPENDIX C

Tail-Up Analysis

Methodology

The F-22A contract contains a separate CLIN for "tail-up"—an increase in manufacturing hours at the end of a production run due to various inefficiencies in fabrication and assembly that result indirectly from production shutdown. Unfortunately, no standard methodology exists for estimating these hours or the price paid for them.[1] We devised two simple methods: (1) direct comparison of hours or dollars in the last lot with previous lots, and (2) improvement curve fitting. We will compare the contracted value for tail-up with these estimates.

The first method was further refined: Last lot average hours were compared with the lot-quantity weighted average hours in (a) the three lots before the last (abbreviated as LL-3), (b) the two lots before last (LL-2), and (c) the next-to-last lot (LL-1).[2] The second method fits a learning curve to all full-rate production lots except the last, and compares the hours in the last lot with the hours forecasted for the last lot from the regression.[3] Identical methods were applied to price data; for

[1] Unfortunately it was not possible to examine how other aircraft contracts have estimated and negotiated tail-up.

[2] The percentage was calculated as the hours in the last lot divided by the weighted average hours, minus 1: $[(hours_{LL}/hours_{weighted\_average}) - 1]$. Identical calculations were performed for dollars.

[3] The improvement curve, ln(Hours) = a + b*ln(Lot Midpoint), regresses the natural log of hours on the natural log of the lot midpoints, estimating coefficients *a* and *b*. Once these are determined, the lot midpoint of the last lot is inputted, and the hours are estimated.

dollars, the learning curve (LC) becomes a cost improvement curve (CIC).

Data

The two sets of annual unit average lot data were (1) manufacturing hours provided by Lockheed Martin for previous and current analyses, and (2) inflation-adjusted dollars taken from publicly available budget documents and the Navy's HAPCA database. Manufacturing hours were available for the C-5A, C-141A, L-1011, S-3A, F-117, P-3A/B, and F-15A-E; the sum of fabrication and assembly hours were used for F-111. Of those eight programs, annual price data were available for F-15A-E only. Hence, we applied the same methodology to the F-15A-E and nine other military aircraft programs. Hours were adjusted for production rate effects using a 97 percent production rate slope; dollars were adjusted using an 89 percent production rate slope.[4]

Analysis

Tail-up percentages for hours are presented in Table C.1; depending on method, average tail-up is estimated at 2.6 percent, 6.1 percent, 7.4 percent, and 10.9 percent. Looking at each program individually yields no programs with a consensus across every method. We do not have enough information to explain the negative tail-up for the F-15 and F-117 compared with the previous lots. Comparing methods across programs, all the results were extremely variable, yet the methods are distinguishable: LL-3 had the smallest tail-up in half the programs, and never the largest; the learning curve has the smallest once, and the largest in six of eight.

---

[4] Lot *n* hours were multiplied by the ratio: $PR_{avg}^{\ln(0.97)/\ln(2)} / PR_{n}^{\ln(0.97)/\ln(2)}$, where PR_{avg} is the mean production rate over all lots and PR_n is the production rate in lot n. Dollar calculations were made similarly. Slopes of 0.97 (hours) and 0.89 (dollars) are consistent with previous research; see Table 5.6 of Younossi, 2001.

Table C.1 applies to prime contractor manufacturing hours, but RAND's SYP model requires a tail-up percentage for hours *and* material dollars. Tail-up percentages for dollars are presented in Table C.2; depending on method, the average tail-up for dollars is 6.3 percent, 7.3 percent, 11.4 percent, and 14.6 percent. These are slightly higher, but still in line with, the tail-up for hours in Table C.1, although F-15A-E estimates are considerably different: much higher for the direct comparisons, but much lower for the cost improvement curve.

In both tables, the cost improvement curve method yielded the highest average estimated tail-up percentage, followed by comparison with the next-to-last lot, then the previous two lots, then the previous three. However, this sequence masks large variability within each method.

The trade-offs involved in using one method over another are worth noting briefly. While improvement curve analyses utilize all

Table C.1
Estimated Last-Lot Tail-Up Percentages for Hours

	C-5A	C-141	L-1011	S-3A	F-111	F-117	P-3 A/B	F-15	Mean
LL–3	2.9	6.2	1.3	–0.5	7.4	0.5	4.1	–1.2	2.6
LL–2	9.4	9.4	5.0	1.4	24.5	–0.5	1.1	–1.9	6.1
LL–1	11.0	11.7	4.4	3.0	30.4	–2.1	5.6	–4.9	7.4
LC	9.2	19.5	13.1	8.6	0.8	6.5	7.8	21.4	10.9

Table C.2
Estimated Last-Lot Tail-Up Percentages for Dollars

	A-6E	AV-8B	A-10	EA-6B	E-3A	P-3C	F-14	F-16[a]	KC-10	F-15 A-E	Mean
LL–3	–0.7	–0.9	2.6	18.2	0.6	–14.5	35.9	3.9	–2.5	19.9	6.3
LL–2	6.6	–5.1	6.7	14.4	–3.6	–13.2	39.7	13.8	–5.0	19.0	7.3
LL–1	20.6	13.3	9.2	6.1	2.1	–9.7	17.2	31.3	–1.4	25.3	11.4
CIC	13.7	16.2	2.0	12.3	–2.6	8.2	27.4	34.6	22.2	12.1	14.6

[a] The last full-rate lot of F-16 was considered to be 1990, even though production continued at much smaller rates until 1998.

available lot information, continual technology insertions and manufacturing process changes can make the inflexible statistical form of the curve inappropriate. Simply put, many program histories do not fit well into the improvement curve. And since the aircraft in the last few lots of a program can be essentially different from those at the beginning, it is tempting to use only the last few lots of data. Yet the potential exists that tail-up is also partly exhibited in the next-to-last lot, meaning that LL-1 can be biased downward.

The tail-up value in the F-22A multiyear negotiated contract is 7.4 percent of TPC. This figure is well within the wide range set by the analysis of various estimating methodologies evaluated in this section. Hence, we chose to use 7.4 percent as a tail-up factor in our SYP model.

APPENDIX D

Institute for Defense Analyses Report Summary

In 2006, the Institute for Defense Analyses (IDA) was tasked to perform a business case analysis (BCA) for the F-22A multiyear procurement (MYP) for Lots 7–9 being considered by the F-22A SPO. At the time of the study, the F-22A program cost was limited by Program Budget Decision (PBD) 720, which restricted funding to $10.438 billion over three fiscal years. As a result, IDA examined five scenarios, which are described in Table D.1.

The scenarios varied by whether procurement was single-year (SYP) or MYP; number of aircraft produced; and whether the budget was constrained by PDB 720. Note that Scenario 4b varies from 4a by including two additional aircraft that would be purchased with the multiyear savings.

Table D.1
IDA BCA Scenarios

Scenario	SYP or MYP?	Number of Units	Budget Constrained?
1	SYP	60	No
2	MYP	60	No
3	SYP	56	Yes
4a	MYP	56	Yes
4b	MYP	58	Yes

To conduct the analysis, IDA utilized the Independent Cost Estimate (ICE) model it had developed for a 2005 report.[1] Its analysis was also based on historical multiyear programs—specifically, the F/A-18E/F/G, C-130J/KC-130J, and F-16A/B/C/D. Using updated data on negotiated Forward Pricing Rate Agreements, it updated its ICE model to provide a revised estimate of the SYP costs. It then assessed MYP savings in six separate categories and applied them to the ICE to arrive at an MYP savings estimate. The assessment of MYP savings was based on statistical regression, historical analysis, and IDA judgments. Following this analysis, IDA arrived at a savings percentage and dollar value for the constrained and unconstrained cases, listed in Table D.2.

IDA concluded that the savings percentages in the constrained and unconstrained cases were the same. It estimated 2.6 percent savings on the air vehicle contract (Lockheed Martin and Boeing) and 2.7 percent savings on the engine contract (Pratt & Whitney). These savings percentages differ from those listed in the table because some of the program funding would not be included in the multiyear contract. Additionally, IDA concluded that the savings generated in the constrained MYP case (Scenario 4a) could purchase two additional aircraft, creating Scenario 4b. This option was also analyzed to assess its procurement costs.

Table D.2
Savings Summary

Budget	Units Purchased	Savings (% of program funding)	Savings ($millions)
Unconstrained	60	2.2	$235
Constrained	56	2.2	$225

[1] Nelson et al., 2006.

IDA Savings Estimate

In estimating multiyear savings for the F-22A, IDA assessed savings in five categories:

- airframe and subsystem suppliers
- avionics suppliers
- labor
- administrative
- propulsion.

It also provided savings in a sixth category, "Below Flyaway," but offered no explanation of methods.

In its report, IDA provided a breakdown for its savings analysis of the budget-constrained case but not the unconstrained (60 aircraft) case. To arrive at a savings breakdown for a 60 aircraft MYP, the airframe and subsystem, avionics, labor, and propulsion savings were assumed to scale proportionately while the administrative and Below Flyaway savings were held constant.

Table D.3
Comparison of Contract Savings

Element	IDA Estimate (All values in $millions)
Airframe/subsystems	68
Avionics	64
Labor	38
Administrative	14
Engine	32
Below the line (BTL) (PSO + PALS + engine BTL)	19
TOTAL	235

NOTE: PSO = Product support, other; PALS = Program agile logistics support.

Airframe and Subsystem Supplier Savings

The airframe and subsystem supplier savings were calculated using the following formula:

$$Savings\ \% = 3.3\ (Lots - 1)^{.53} .52^{(Structure)} .45^{(Boeing)} + 5.9\ (EOQ\%)$$

Structure and *Boeing* are binary dummy variables indicating if a given item is structural and/or produced by Boeing, respectively; *Lots* is the number of lots in the multiyear contract (in this case, 3); and *EOQ%* is the percentage of funding coming from EOQ. IDA's formula was derived from a least-squares regression of supplier responses and multiyear procurement estimates in AA05[2] data. IDA divided the data between Boeing and Lockheed suppliers and between "structural" and "systems" components, excluding avionics (which were estimated in a separate section). The regression appears to have been conducted for each category, rather than each system, in order to capture savings for systems without any data.

Avionics Savings

IDA's avionics savings were assessed from estimates provided by the Big Four manufacturers in the AA03 data, a second analysis from 2005, a 2005 Lockheed analysis of supplier submissions, and data collected by IDA. From these data, IDA determined a savings rate for each system and averaged them into an overall savings rate for avionics, weighted by contract value. IDA claimed that the Big Four accounted for 80 percent of the avionics price.

---

[2] AA05 is the affordability assessment produced in 2005. The USAF required that the contractors provide estimates every other year for several years. AA05 was the last one produced for the F-22.

Labor Savings

No labor savings were assessed in AA05, but IDA determined that there should be some, based on a historical example. They divided these savings into supplier support labor and TPC/PSAS engineering hours. Support labor savings were assessed at 3.3 percent, while engineering savings were assessed at 5 percent, based on a historical analysis of the F/A-18 program.

Administrative Savings

Administrative savings were based on Lockheed Martin and Boeing assessments conducted for a five-lot multiyear contract as part of AA05. The assessments included savings from eliminating affordability assessment estimates, which IDA removed. In addition, IDA eliminated savings from proposal preparation for Lot 7, as the Air Force was requiring an SYP proposal for that lot, and for Lot 10, which would not be included in a multiyear.

Propulsion

IDA's propulsion savings were based on the following regression, derived from several historical programs:

$$Savings\ \% = .35 * PoP + .01175 * QTY$$

where *Savings %* is the savings percentage; *PoP* is period of performance for the contract (i.e., number of fiscal years in an MYP); and *QTY* is quantity of engines produced.

Below Flyaway

IDA also included "Below Flyaway" savings, which included PALS and PSO contracts.

APPENDIX E

Military Fixed-Wing MYP Since 1995

This appendix reviews the seven more-recent MYP contracts since 1995 in somewhat greater detail than described in Chapter Five. Contracts that have one or more key characteristics similar to the proposed F-22A MYP contract are discussed in somewhat greater depth.

C-17 MYP I

The Boeing (formerly McDonnell Douglas)[1] C-17 Globemaster III is the USAF's premier strategic airlifter. In May 1996, the USAF and McDonnell Douglas signed a $14.2 billion MYP contract for 80 aircraft to be procured over seven years (FYs 1997–2003). This was the largest and longest multiyear contract ever negotiated up to this point. According to the original proposal, the contractor and the Air Force estimated that the MYP contract with McDonnell Douglas for 80 aircraft would save 5.5 percent or about $900 million (all in TY$) compared with a series of SYP contracts over a comparable period for the same number of aircraft.[2] At the same time, the Air Force also signed

---

[1] Boeing announced the planned acquisition of McDonnell Douglas for $13.3 billion in December 1996.

[2] See Congressional Research Service, 2000, p. 3. There is some uncertainty from the available data as to whether the final savings estimate shown above for the aircraft included multiyear savings on the engine program. This is because, unlike all other fixed-wing multiyear programs after 1995, no formal multiyear justification for Congress was published at the time of program initiation, because the program was proposed outside the normal budget

a multiyear contract with Pratt & Whitney for the commercial derivative engine program (CDE) for procurement of the F117 engine to power the C-17. This contract was valued at $1.6 billion; the Air Force estimated the multiyear procurement program for the engine saved $88 million or 5.5 percent over SYP.[3]

By early 1994, it had become evident that the C-17 program, which at that time was in the low-rate initial production (LRIP) phase (about 30 delivered or in assembly), was experiencing significant cost growth, schedule slippage, and quality and performance problems.[4] The relationships between the prime contractor, the Air Force Program Office, and the Defense Contract Management Command (DCMC) had become intensely adversarial. At this point, the contractor and the USAF had a strong motivation to fix the program, as discussed in greater detail below.

Four separate (but partially overlapping) "get well" initiatives were launched for the C-17 at this time:

1. "Should Cost," August 1994–February 1995
2. "Lot VIII & Beyond," January 1995–August 1995

---

cycle. According to a contemporary GAO report published before the contract negotiation with McDonnell, the original total program savings estimate was for 5 percent, or TY $896 million. This account claims the estimated savings came primarily from two sources: the airframe contract and the engine contract. According to this account, McDonnell Douglas reduced its contract price for 80 aircraft by about $760 million, or 5 percent. The USAF expected to realize a 6 percent savings, or approximately $122 million, on the engine contract with Pratt & Whitney. The Air Force estimated that the remaining savings of about $14 million would come from other sources. See GAO/T-NSIAD-96-137, 1996, p. 6. However, other sources seem to indicate that the $900 million estimate of savings in the final contract with McDonnell did not include the engines. For example, see "C-17 Globemaster III Production," at GlobalSecurity.org.

[3] Beginning with the Lot 4 buy in November 1992, the government took over procurement of the F117 engine and provided it to McDonnell Douglas as government furnished equipment (GFE). See DoD News Release, 1996.

[4] This introductory account is drawn from open sources and from notes on a Boeing briefing entitled "Doing Business Differently" presented by the Boeing Production Pricing Group to a group of RAND analysts at Boeing Long Beach on October 7, 1999.

3. "Price Based Acquisition" and "Must Cost" commercialization initiatives
4. "Multi-Year Procurement," January 1996–May 1996.

Before the MYP contract, the USAF and Boeing focused on two areas to save costs covered in the first three initiatives listed above: increasing manufacturing/assembly efficiency and enhancing the product design for producibility. The emphasis in the first and second initiatives was on reducing indirect costs, outsourcing (transferring risk and cost responsibility to suppliers), and implementing the Lean Aircraft Initiative concepts. In the third initiative, the focus was on upgrading avionics, inserting commercial microcircuits and other parts components, and designing for manufacturing and assembly (DFMA). In late 1995, the parties moved to a fourth strategy of MYP to reduce price by using up-front funding for gaining economies of scale on procurement of materials and to fund additional affordability projects.

The MYP initiative was stimulated by a Defense Acquisition Board (DAB) Acquisition Decision Memorandum for an option for an accelerated production profile, which would reduce the 80 aircraft average price to $188 million. Air Force leadership determined that a contract savings percentage of greater than 5 percent had to be demonstrated to justify an MYP on a program that was so troubled politically. The contractor responded with a "Management Challenge," promising a 5.5 percent further reduction to $178 million average aircraft cost in return for a multiyear procurement commitment for 80 aircraft, resulting in a total contract value for 80 aircraft of $14.2 billion.

The government accepted, and the contract award for a seven-year multiyear contract with Boeing for 80 aircraft, valued at $14.2 billion, became effective in June 1996. The government provided $300 million in advance for EOQ (parts and material) and additional affordability projects, in anticipation of receiving a payback ratio of 3:1 or $900 million in savings. Affordability projects and supplier material savings were identified, and the estimated savings were put on contract.

EOQ Funding

The C-17 MYP I specified $300 million in government EOQ funding, which could be and was spent on CRIs. This was on top of extensive additional government and industry CRI funding immediately preceding and throughout MYP I (see below). The formal EOQ funding amounted to less than 2 percent of the initial forecast contract value. According to some sources, the Air Force assumed that McDonnell Douglas would spend about one-third or $100 million of the EOQ funding on nonrecurring CRIs.

Phasing of EOQ Funding

Budget data on the phasing of the EOQ funding and other sources indicate that the EOQ funding was initially made available in FY 1996 before the formal beginning of the MYP contract, indicating that this money or much of it was available from the very beginning of the MYP contract, for economic purchases from vendors and for CRIs. "Payback" of the EOQ funding was spread out over the seven-year MYP contract.

Annual and Total Procurement Numbers

This MYP contract entailed relatively modest procurement numbers, with a total of 80 aircraft over seven years. The actual annual procurement contract numbers from FYs 1997 through 2003 were as follows: 8, 9, 13, 15, 12, 15, 8. This varied slightly from but remained close to the original proposal.[5]

Duration

C-17 MYP I lasted seven years, making it the longest fixed-wing military aircraft MYP contract we are aware of. The GAO noted at the time that a seven-year MYP contract was unique and that the current statutory limit was five years.[6]

---

[5] The original proposal had called for procurement of 15 aircraft in FY 2001 Lot 13, rather than 12, and 5 aircraft in FY 2003 Lot 15 instead of 8.

[6] GAO/T-NSIAD-96-137, 1996, p. 3.

CRI Funding

The C-17 MYP I enjoyed considerable CRI funding. EOQ funding could be used for CRI. Initially the Air Force anticipated that about one-third of the $300 million, or $100 million, would be used for CRIs. The GAO reported that early in the program McDonnell Douglas had trouble identifying sufficient numbers of EOQ opportunities and was achieving disappointing results from discussions with its vendors.[7] CRI projects funded with MYP EOQ funding included development of a composite horizontal stabilizer, overhead panel redesign, the Electrical Product Improvement Program (EPIP), and many others.

Significant additional CRI funding that overlapped with the MYP contract were made available to the prime contractor through separate contracts. The Omnibus CRI Program, part of the January 1994 Settlement Agreement that set the stage for the MYP contract, called for no less than $100 million to be invested by the contractor in CRIs. The Omnibus program allowed for CRI investments in FYs 1996 and 1997 during the MYP contract. In July 1995, the Air Force awarded a new and separate CPAF contract to McDonnell for a variety of CRIs that overlapped the MYP contract. These included the Affordability 95 and Nacelle/Engine Affordability Team (NEAT) programs for a total of $295 million in Producibility Enhancement/Performance Improvement (PE/PI) programs, which required a 3:1 return on government investment.[8] This was followed in 1999 by the "Must Cost" CRI program aimed at reducing production costs beyond aircraft 120. Originally intended to entail $275 million in funding provided by Boeing from profits and then paid back by the government, this program eventually was pared back to $212 million. Finally, in 2001 and 2002 Boeing invested money in further CRIs in anticipation of a follow-on MYP contract. This program was called Affordability 180.

In total, approximately $560 million in CRI money was invested during or at least partly overlapped the MYP program period (excluding Affordability 180), nearly all of which was provided by government contracts separate from the MYP contract. In addition, it appears that

---

7 GAO/T-NSIAD-96-137, 1996, pp. 5–6.

8 GAO/NSIAD-97-38, 1997, p. 14.

at least $100 million of the EOQ funding went to CRIs and probably more. However, it is difficult to determine the precise amount of CRI funding that specifically affected the MYP I period. We have assumed that approximately $300 million is a reasonable figure for MYP I, but this is only an estimate on our part.

Business Base and Industrial Base Environment

As noted above, the C-17 MYP I arose at a time when the program faced stark challenges and possible cancellation. In addition, it was the only major military program under way at McDonnell Douglas's Long Beach facility, at a time when its commercial transport business was in serious decline. The prime contractor therefore was strongly motivated to reduce costs to the government.

Two events in 1994 triggered the contractor and government initiatives to cut costs and motivated the initiation of an MYP program: (1) OUSD(A&T) made it clear that production of the C-17 would end at the current initial lot of 40 unless the program was fixed; and (2) Boeing (which of course was a competitor because it had not yet bought McDonnell Douglas) generated a proposal for a non-development item (NDI) transport aircraft based on the B-747 that would have an average unit production price about half of the projected C-17 price. This Boeing proposal became a formal OSD Defense Acquisition Pilot Program (DAPP) called the Non-Developmental Airlift Aircraft (NDAA). The existence of the NDAA program placed enormous pressure on McDonnell Douglas and the USAF to turn the C-17 program around, and the MYP contract became a key element of that strategy.

Reducing the cost of the C-17 became even more crucial after November 1996, when McDonnell Douglas was eliminated from the Joint Strike Fighter program. Following the Boeing takeover of McDonnell Douglas, the problems encountered by Boeing in its commercial aircraft division and its status as underdog in the JSF competition helped it maintain its incentive to reduce costs on the C-17.

Production Rate Changes

The C-17 MYP I contract included a significant increase in production rates over prior years. The originally projected production rates for FYs 1994 through 1996 were 6–8 aircraft per year. The MYP contract nearly doubled this production rate by raising production from 8 in FY 1997 to 15 per year in FYs 2000 through 2002. Thus, some savings on the MYP contract may be attributable to increased production rates.[9]

Program Maturity

C-17 MYP I took place early in the production program after LRIP, providing numerous opportunities for CRIs.

Foreign Sales

In September 2000, the United Kingdom finalized an agreement to lease four C-17s. All four were delivered during FY 2001, permitting an increase in planned production for FY 2003 from five aircraft to eight by delaying three USAF C-17s for two years.

C-17 MYP II

The C-17 MYP II program, covering the procurement of 60 aircraft over five fiscal years (FYs 2003–2007), claimed a $1.3 billion cost savings over an estimated annual procurement cost of $12.8 billion, for an overall 10 percent cost savings for the airframe and engine contracts. The Air Force estimated a 10.8 percent cost savings on the airframe procurement contract with Boeing over annual procurements (or $1.211 billion) and 5.7 percent cost savings on the Pratt & Whitney engine contract (or $92 million). More specifically, the Air Force estimated the annual cost of an SYP program at $12.805 billion versus

[9] However, it appears that before the MYP was proposed, McDonnell offered to lower average unit airframe price from $208 million to $188 million, at least in part due to the newly proposed higher annual buy profile. Sparks, 2004.

$11.503 billion for an MYP program.[10] In August 2002, the Air Force awarded Boeing a $9.7 billion MYP contract for 60 C-17 airframes.

The C-17 MYP II was based on an unsolicited proposal using Price Based Acquisition (PBA) without cost data, and assumed an FFP contract. The proposal was only nine pages long. The build proposal assumed the maintenance of a 15-aircraft-per-year production rate through FY 2006, by adding an additional seven aircraft to FY 2003. FY 2007 (Lot 19) included eight aircraft, for a total of 60.

MYP II was initially intended to be a commercial-like PBA contract award under FAR Part 12 commercial item contracting rules. Failure to receive a commercial item determination led to the award of a FAR Part 15 traditional FFP contract with waivers. The Air Force received a Truth In Negotiations Act (TINA) waiver and waived the contractor requirement for Contractor Cost Data Reporting (CCDR) because of the extensive history of C-17 cost data and the existence of the SPO-contractor C-17 Joint Cost Model (JCM). The contract was negotiated using price data through Lot 12 (FY 2000), but no detailed contractor price data were received by the USAF for MYP II. Boeing was awarded an FFP contract at $9.762 billion for 60 aircraft in August 2002.[11]

This MYP contract raised some concerns in Congress and elsewhere because the planned Air Force procurement quantity for FYs 2004 and 2005 was less than the planned production rate. The Air Force planned to use Advance Procurement Funding (AP) to help fund the production of aircraft that would be procured in later fiscal years. Thus AP funding was higher than in traditional programs (15 percent of aircraft price on MYP II compared with 11 percent of aircraft price on MYP I, according to the Air Force). As a result, the Air Force cancellation liability ceiling was much higher (at around $1.5 billion at its height) than on the C-17 MYP I because the contractor was building aircraft not yet purchased and his forward exposure had to be covered. The cancellation ceiling was not funded.

---

[10] DoD, *C-17,* Exhibit MYP-1, Multiyear Procurement Criteria, February 24, 2003a, pp. 2–3.

[11] Congressional Research Service Report to Congress, 2006.

No government-funded or separately contracted CRIs were developed specifically for C-17 MYP II. Furthermore, with no cost reporting, the Air Force had minimal insight into how the contractor spent the AP and EOQ funds. The Air Force believed the bulk of the savings came about for three reasons: (1) the Air Force provided Boeing with a more favorable cash flow; (2) Boeing received price reductions from suppliers by contracting at more economic quantities; (3) and AP funding and progress payments permitted Boeing to maintain overall production at the most efficient rate, even when procurement fell below that rate.

EOQ Funding

The 2003 justification projected $645.2 million in EOQ funding over the course of the MYP contract, in addition to $1.098 billion in AP money. Some Air Force sources seem to suggest that at least some of this EOQ funding was used for maintaining Boeing's cash flow and the annual build rate of 15 aircraft per year when procurement fell below that number. According to Boeing officials, however, significant additional savings were negotiated as discounts with many vendors and suppliers on the order of 10–20 percent, because of the MYP contract. Whether these were funded with EOQ money is not known.

Phasing of EOQ Funding

According to the February 2003 justification, EOQ funding was heavily front-loaded, beginning in CY 2002 at $180.9 million, peaking in CY 2003 at $211.8 million, and ending in CY 2005.

Annual and Total Procurement Numbers

Even more than the C-17 MYP I contract, the MYP II contract entailed relatively modest procurement numbers, with a total of 60 aircraft over five fiscal years. As noted above, the contractor built at the preferred economic rate of 15 aircraft per year, while the Air Force back-loaded procurement, with the difference covered by AP funding. The originally proposed buy profile covered six fiscal years and included 4 aircraft in FY 2003 followed in successive fiscal years by 10, 11, 14, 16, and 5. This was later adjusted to 7 in FY 2003, followed by 11, 14, 15, and 13. Still later, under congressional pressure, this was again adjusted and more closely aligned with production rates as follows (beginning with FY 2003): 7, 11, 15, 15, and 12.

Duration

The C-17 MYP II was a typical MYP contract lasting five fiscal years, although originally the Air Force had requested (and been denied) a six-year MYP contract for the same total number of aircraft.

CRI Funding

No formal government-funded CRI projects were contracted for on the C-17 MYP II. However, the 1999 $212 million government-funded "Must Cost" initiative and the Boeing-funded "Affordability 180" program in 2001–2002 were aimed at reducing costs for a second MYP contract, which of course became MYP II. In addition, since the SPO had little insight into how Boeing spent EOQ money in MYP II, some of this may have been spent on CRIs.

Business Base and Industrial Base Environment

With the commercial aircraft downturn after September 11, 2001, and Boeing's loss to Lockheed in the final selection of the JSF program in October 2001, the contractor was highly motivated to reduce costs on the C-17 program, and maintain economic production rates for the F/A-18E/F (see below).

Production Rate Changes

As noted above, the Air Force and contractor went to great lengths, including creative financing approaches, to maintain the economic build rate of 15 aircraft per year achieved in MYP I, even though full procurement funding was not available to sustain this rate early in the program.

Program Maturity

The C-17 MYP II took place during the more mature production phases of the program, following the production of 120 aircraft. However, Boeing had reasonable expectations for follow-on procurement after the end of MYP II, since the MYP II contract included a possible increase of 42 aircraft after the 60 MYP II aircraft. Indeed, in October 2006, Congress provided $2 billion in unrequested funding to procure an additional ten aircraft in FY 2008.[12]

Foreign Sales

Additional foreign sales may be emerging. The United Kingdom allegedly plans to purchase outright 45 C-17s. Both Australia and Canada have purchased four C-17s each. NATO may also purchase three to four. These orders will likely be filled after the conclusion of the MYP II contract.

E-2C MYP I

The E-2C Hawkeye is a U.S. Navy carrier-based tactical airborne warning and control system platform. The first E-2C MYP contract covered FYs 1999–2003. It was a single five-year firm-fixed-price contract for the airframe only. The original Navy MYP justification (February 1998) showed 8.3 percent total airframe contract savings or $106.5 million over annual contracts with the same quantity profile, with the total MYP airframe contract price originally estimated at $1,181.3 million. The total procured quantity was very low: 21 aircraft. There was a relatively large amount of EOQ funding for material for 21 shipsets of "detail parts," and contracting for 21 ship sets of "Prime Mission Equipment" was provided during one lot buy in FY 1999. Indeed, over one-third of the contract value was EOQ funding. GFE included engines and Joint Tactical Information Distribution System (JTIDS). CFE included the radar, Passive Detection System (PDS), rotodome, landing gear, Identification Friend/Foe (IFF), and other equipment.

In April 1999, Northrop-Grumman was awarded a $1.3 billion five-year MYP contract covering 22 Hawkeye 2000 (upgraded E-2Cs), which included 21 for the U.S. Navy and one for the French Navy. Later, two more foreign military sales aircraft were added.

---

12 Congressional Research Service Report to Congress, January 25, 2007.

EOQ Funding

This MYP had a relatively large amount of EOQ funding. AP funding, mostly made up of EOQ funding, was projected at $417.9 million. With a total airframe contract price originally estimated at $1181.3 million, this accounted for approximately 35 percent of total contract price. NAVAIR claims the vast bulk of savings came from EOQ funding and negotiated quantity discounts from suppliers and vendors.

Phasing of EOQ Funding

EOQ funding was allocated for purchases applicable to at least four of the five years of the MYP, and was heavily front-loaded for FYs 1999 and 2000.

Annual and Total Procurement Numbers

The E-2C MYP I had very small annual and total procurement contract numbers. Total procurement was originally planned at 21 procured beginning in FY 1999 as follows: 3, 3, 5, 5, 5.

Duration

This MYP contract covered a typical five-year duration.

CRI Funding

No apparent explicit CRIs were funded or called out in the contract or justification. NAVAIR claims that no AP or EOQ funding was formally spent on CRIs.

Business Base and Industrial Base Environment

Uncertain, but business base was probably declining because of the approaching conclusion of A/EA-6F production and other work.

Production Rate Changes

The MYP contract envisioned a slight rate production decrease then a slight rate increase compared with immediately prior years. The SYP contracts for FYs 1995 through 1998 were for 4, 3, 4, and 4, respectively.

Foreign Sales

There were three foreign sales aircraft during the MYP: one to France and two to Taiwan, raising the total production from 21 to 24.

E-2C MYP II

The E-2C MYP II contract in many respects does not constitute a true MYP, according to NAVAIR. It is the smallest fixed-wing military MYP contract since 1995. Significant savings were not claimed to be the primary motivation for the MYP contract. Rather it was an attempt to fill a production gap and keep the production line warm between the end of E-2C production and the beginning of LRIP for the significantly upgraded E-2 Advanced Hawkeye, as it was called at the time. However, the Navy's formal justification did claim a cost savings over annual procurement of 7.2 percent for the airframe and engine procurements. The E-2C MYP II consisted of two four-year, fixed-price contracts, one for the engines and one for the aircraft, covering FYs 2004–2007. The entire procurement consisted of four E-2C aircraft and four TE-2C aircraft and 16 engines. The total MYP procurement contract price (airframe and engine) was estimated in the justification at $788.6 million, compared with an annual total contract price estimated at $850.0, for an estimated cost savings of $61.5 million for the airframe and engine contracts.

EOQ Funding

Like the E-2C MYP I, this MYP also had a relatively large amount of EOQ funding, with an estimated $85.8 million in EOQ funding. Thus, EOQ funding made up over 10 percent of the estimated contract price.

Phasing of EOQ Funding

As in the case of the first MYP, the E-2C MYP II EOQ funding was allocated for purchases applicable to at least four of the five years of the MYP, and was heavily front-loaded for FYs 2003–2005.

Annual and Total Procurement Numbers

The E-2C MYP II had extremely small annual and total contract numbers. Total procurement was originally planned at eight aircraft, with two procured each year from FYs 2004 through 2007.

Duration

This MYP contract covered a shorter-than-usual four-year period and appears to be the shortest fixed-wing MYP program to date in the 1995–2007 time frame.

CRI Funding

As in the case of MYP I, no apparent explicit CRIs were funded or called out in the contract or justification.

Business Base and Industrial Base Environment

Both the Navy and Northrop-Grumman were strongly motivated to keep a minimum level of production going at Northrop-Grumman St. Augustine to bridge the period between the end of the E-2C 2000 and the beginning of LRIP for the newer upgraded advanced variant initially called E-2 Advanced Hawkeye.

Production Rate Changes

The MYP contract envisioned a significant production rate reduction from five aircraft to two aircraft per year.

Foreign Sales

Several existing E-2C foreign customers signed contracts to upgrade their aircraft to E-2C 2000 configuration, as the version under production in the MYP II was eventually called.

CC-130J (USAF) and KC-130J (USMC) MYP

In March 2003, the Air Force awarded Lockheed Martin a $4.05 billion six-year joint Air Force/Marine MYP contract for procurement of 60 CC-130J and KC-130J Super Hercules aircraft from FYs 2003 through 2008.[13] This includes 40 CC-130Js, a stretched version of the C-130J tactical airlifter being procured by the U.S. Air Force, and 20 KC-130Js, an aerial tanker/transport version of the C-130J procured by the U.S. Marine Corps. As of mid-2006, total annual deliveries of both types combined were scheduled as follows: 4, 4, 15, 13, 13, 11.[14] MYP total airframe contract savings compared with SYP were originally estimated at 10.9 percent or $513.07 million.

This is a fixed-price contract with production/quantity rate change adjustment factors. Originally this was a commercial item FAR Part 12 Price Based Acquisition contract (FFP + EPA) with no formal cost and price reporting. This contract is currently being restructured as a traditional FAR Part 15 contract.

The contract renegotiation from FAR Part 12 to Part 15 was announced in October 2006, and applied to the last three fiscal years or 39 aircraft. According to an Air Force Print News report, the Air Force estimated that an additional "institutional net savings" of $168 million above the original MYP contract savings would be realized due to the negotiated repricing of the last 39 aircraft using TINA compliant cost and pricing data from the contractor.

According to the original MYP justification, $140 million in EOQ funding was planned for FYs 2003–2005, and $540 million of advance procurement funding for FYs 2004–2007.

The original cost estimates and claimed MYP savings were based on contractor draft proposals for MYP and annual procurement of 62 aircraft, as well as SPO estimates based on experience with annual procurements during FYs 1994–2002. However, prior to FY 2004, all lots were bought under commercial FAR Part 12 contracts, so Lockheed Martin did not report detailed cost and pricing data that the govern-

---

[13] DoD, Office of the Inspector General, 2006, p. 4.

[14] The original MYP justification envisioned a 62-aircraft procurement with total annual procurement beginning in FY 2003 as follows: 12, 13, 12, 13, and 12. Forty-two USAF CC-130Js are shown on the Contract Funding Plan as a five-year MYP with five procured in FY 2004 and the remaining 38 procured from FYs 2005 to 2008. KC-130J Marine aircraft are shown procured as follows: four in FY 2003, and four each year from FY 2005 to FY 2008, for a total of 20 aircraft.

ment could use to make detailed estimates of SYP versus MYP contracts for the final contract negotiations.

EOQ Funding

The C-130J MYP permits multiyear buys from vendors that provide cost savings. EOQ funding is a relatively modest but significant component of the total contract value. At $140 million, this amounts to about 3.5 percent of the original MYP contract value. Advance procurement amounts to about 14 percent of the MYP contract. EOQ plus AP is stated as a key driver of cost savings in the original justification. No advance procurement (AP) is shown for the SYP comparison for calculating MYP savings in the justification.

Vendor-supplied items are identified as making up more than 50 percent of aircraft value. According to acquisition officials, suppliers were permitted to buy out items for the entire procurement contract under a single contract order, or continue producing at the most economical rate. Acquisition officials claim that Lockheed Martin successfully negotiated supplier prices down about 13 percent on average for a FY 2003–2008 buy.

However, acquisition officials noted that before the switch from FAR Part 12 to FAR Part 15 contracting, the government had limited insight into how the prime contractor spent EOQ funding. Lockheed Martin could determine how best to spend its EOQ funding, and apparently at least some of it was spent on CRIs.

Phasing of EOQ Funding

According to the original program funding plan, the EOQ and AP funding would begin in FY 2003 at a low level (covering only four KC-130Js) then continue throughout the rest of the MYP program (to FY 2007) at roughly equal annual amounts.

Annual and Total Procurement Numbers

The C-130J MYP contract covers a relatively small total quantity of 60 aircraft with small annual procurements ranging from 4 to 15 aircraft.

Duration

The C-130J MYP is an unusual joint six-year MYP contract covering FYs 2003–2008. However, the Air Force component is a more standard five-year MYP within the joint program.

CRI Funding

No clear indications of significant new CRI spending for the MYP program were easily identifiable. No specific CRI money was called out in the original February 2003 MYP contract justification. As a Price Based Acquisition with an FFP under FAR Part 12 commercial item rules, there was little visibility into Lockheed Martin's costs and how money was spent. Some EOQ money was likely spent on CRIs. Furthermore, the MYP justification claimed that the prime contractor was likely to make process improvements because of the MYP.

Business Base and Industrial Base Environment

The C-130J was originally developed with a significant amount of contractor money and offered to the Air Force on a commercial type PBA basis, which required significant production quantities to recoup Lockheed's nonrecurring investments. Before the beginning of F-22A LRIP production in 2000, the C-130J was the only major full-rate production program under way at the Lockheed plant in Marietta, Georgia. In December 2004, DoD Program Budget Decision 753 proposed terminating procurement of the C-130J after FY 2005 for the Air Force and after FY 2006 for the Marine Corps, reducing the MYP contract from 60 to 25 aircraft. In 2005, congressional pressure combined with concerns over termination liability costs prevented termination of the program. However, these and other pressures led to the 2006 renegotiation of the contract under traditional FAR Part 15 rules, providing a reported additional 8 percent reduction in price for the remaining 26 C-130Js for the Air Force.[15]

---

[15] DoD, Office of the Inspector General, 2006, pp. 1–7; and "C-130J Acquisition Program Restructured," November 9, 2006.

Production Rate Changes

Annual procurement rates varied widely prior to the MYP. In October 1995, the USAF contracted for two C-130Js in a modification of the C-130H contract. Production began in 1996. A November 1996 contract was a five-year option contract for 35 aircraft. In December 2000, a second five-year option contract was signed for 20 aircraft. By December 2003, 50 out of the 57 aircraft on order by the U.S. government had been delivered: 14 in 1999; 6 in 2000; 13 in 2001; 8 in 2002, and 9 in 2003. However, significant foreign orders were also received during this period.

Foreign Sales

Prior to the MYP, several countries ordered small numbers of the C-130J, including Australia, the UK, Denmark, and Italy. In November 2006, press reports indicated that Canada had agreed to a $4.4 billion contract for 17 C-130Js.

F/A-18E/F MYP I

In June 2000, the U.S. Navy launched full-rate production of the F/A-18E/F Super Hornet fighter by signing a five-year MYP contract (FYs 2000–2004) with Boeing for $8.9 billion covering the procurement of 222 aircraft (later reduced to 210). The original justification documentation claimed a cost savings over annual contracts of TY $706 million or 7.4 percent on the total airframe contract price with Boeing. The contract was an FPI-type contract, with a 70:30 split.

An additional five-year MYP contract covering FYs 2002–2006 was proposed in 2002, with estimated savings of 2.8 percent. In July 2002, the U.S. Navy awarded GE a $1.9 billion five-year MYP contract for 480 F414 engines, devices, and spare modules.

According the F/A-18E/F program office, the most important sources of savings on the main airframe MYP contract with Boeing were the $200 million in EOQ and CRI funding and annual AP funding. Significant price reductions from suppliers were achieved from EOQ funding, but, according to the SPO, CRI funding ($115 million) was a much more important source of cost savings than EOQ funding and provided a higher return on investment.

EOQ Funding

The Boeing airframe contract contained a relatively small amount of EOQ funding, amounting to $85 million or about 1 percent of the contract. The original justification shows vendor procurement as the largest source of estimated cost savings, accounting for about 68 percent of the total anticipated savings. However, much of this was due to quantity discounts derived from the large total procurement numbers and from CRIs, rather than specifically from EOQ funding. According to NAVAIR, there was only a limited use of EOQ funding, and it was found to be in some respects divergent from CRI funding objectives. Apparently at least some and possibly the majority of EOQ funding was used as nonrecurring funding for CRIs.

Examples of where traditional EOQ funding produced savings are in the areas of raw materials and forgings, but only over a two fiscal year period. This was because the time value of money did not justify more than two years. Savings directly attributable to EOQ, presumably realized through discounts from selected suppliers, were estimated by NAVAIR to be in the 5–10 percent range.

Phasing of EOQ Funding

EOQ funding was made available in roughly equal amounts from FYs 2001 to 2004, with a slight bias toward the later fiscal years.

Annual and Total Procurement Numbers

The F/A-18EF MYP I originally envisioned the procurement of 222 aircraft (later reduced to 210 aircraft), a very large number compared with many other MYP contracts. These large procurement numbers contributed significantly to the ability of the prime contractors to negotiate vendor price reductions and achieve other efficiencies of scale, according to NAVAIR. The annual production rate varied from 36 to 48 aircraft per year. Beginning in FY 2000, annual production rates were: 36, 39, 48, 45, and 42.

Duration

The F/A-18E/F MYP I was a typical five-year MYP spanning FYs 2000–2004.

CRI Funding

CRI funding of $115 million was provided in the F/A-18E/F MYP I contract. Like EOQ, this was also roughly 1 percent of the total contract. Unlike EOQ, however, the CRI funding was heavily front-loaded, with about 77 percent made available in FY 2000 in the first year of the MYP contract. According to NAVAIR, CRI funding was the principal source of savings in MYP I and had been emphasized to a much greater extent than EOQ efforts, because savings potential from CRI appeared to be much higher. CRI opportunities were evaluated on a case-by-case basis, and CRI money flowed down to vendors when major returns on investment seemed possible. CRIs included process improvements, design improvements, increased automation, make/buy decisions, and so forth. Overall, CRIs had to demonstrate a better than 3:1 return on investment to be pursued, according to NAVAIR, which was a much better return than experienced on EOQ funding.

Business Base and Industrial Base Environment

NAVAIR argued that every MYP is unique. The ability to achieve cost savings is driven to a substantial degree by the business environment at the time the MYP is negotiated and by the business base of the prime contractor and major vendors. The business base environment for Boeing at this time was similar to that discussed above for the C-17 MYP II.

Production Rate Changes

MYP II followed three LRIP lots for a total of 62 aircraft covering FYs 1997 through 1999. Total production for FY 1999 was for 30 aircraft, following annual production of 20 and 12 in FYs 1998 and 1997, respectively. Thus MYP I represented a significant increase in production rate up to 45 aircraft a year in FY 2003. Production rate increases generally help reduce unit costs.

Program Maturity

The F/A-18E/F MYP I transitioned the program from LRIP to full-rate production. Thus the MYP contract was negotiated in the earliest phases of production. This may help explain why there seemed to be so many high-leverage CRI opportunities available to the program.

Foreign Sales

There were no foreign sales of F/A-18E/Fs during MYP I, but total U.S. Navy and Marine procurement was so high that foreign sales were relatively less important for this program.

F/A-18E/F MYP II

In December 2003, the U.S. Navy awarded a second five-year MYP contract to Boeing valued at $8.9 billion. The contract originally envisioned procurement of 222 aircraft (later reduced to 210), made up of 154 F/A-18E/F aircraft and 56 EA-18G electronic attack aircraft (which has an airframe identical to F/A-18E/F). The contract spans FYs 2005 through 2009. According to the February 2003 justification, the proposed MYP contract includes a cost savings of $1.052 billion, or an estimated 10.95 percent savings over single-year procurements estimated to total $9.612 billion. According to NAVAIR, the savings for MYP II represent in essence a 10.95 percent drop in average unit price from the last MYP I lot, with the price essentially remaining unchanged from that point on (in constant FY 2000 dollars).

Unlike MYP I, which was an FPI contract with a 70:30 split share, the MYP II contract is FFP using a PBA strategy, with TINA waivers and with minimal cost reporting and cost/pricing insight. The intention was a full transfer of the responsibility for realizing CRIs and the target price to the contractor. The incentive for the contractor to lower costs was that any underruns would all be retained by the contractor.

According to the justification and NAVAIR, all contract cost savings were expected to come from CRIs, most of which were identified during the ongoing "Must Cost" initiative undertaken during MYP I.

The contract provided $100 million investment funding for CRIs, but no EOQ funding whatsoever.

EOQ Funding

Unlike most recent MYP contracts, no EOQ funding was sought for the F/A-18E/F MYP II, and consequently the contract includes no EOQ funding at all. According to Navy procurement officials, this strategy was based on a review of lessons learned on the sources of cost savings in MYP I conducted early in 2002, which indicated that CRIs were the principal source of cost savings on MYP I.

Phasing of EOQ Funding

Not relevant.

Annual and Total Procurement Numbers

With 210 aircraft, the F/A-18EF MYP II is identical to MYP I in total procurement numbers and is, by recent standards, a very large MYP. The annual production rate is intended to remain constant at 42 aircraft per year, the same procurement number as planned for the last fiscal year of MYP I.

Duration

Like the F/A-18E/F MYP I, the MYP II is a standard five-year MYP spanning FYs 2005–2009. There is no production gap between MYP I and MYP II.

CRI Funding

At $100 million, CRI funding is marginally less than the $115 million provided for MYP I. However, the percentage of the total contract remains roughly the same at about 1 percent. The vast bulk of savings on MYP II are expected to come from CRIs, whether government- or contractor-funded. The CRI funding is aimed at CRI projects identified during MYP I with at least a 3.5:1 return on investment. Most of these CRI savings are expected to be realized on the vendor level. Vendors are being provided with nonrecurring CRI money and are required to provide a payback based on a negotiated settlement.

The February 2003 program justification estimates that 66 percent of the savings would come from vendor procurement. According to NAVAIR, a later estimate showed that 77 percent of the nonrecurring CRI money went to vendors and suppliers, who provided about 69 percent of expected savings.

Business Base and Industrial Base Environment

As in the MYP I case, NAVAIR argued that every MYP is unique and that the ability to achieve cost savings is driven to a substantial degree by the business environment at the time the MYP is negotiated and by the business base of the prime contractor and major vendors. With the C-17 production program nearing its completion at Long Beach; the F-15, T-45, and Harrier production ended or nearing their end in St. Louis; and Boeing's elimination from the JSF/F-35 competition, the F/A-18E/F remained the only fighter production program available and a key component of Boeing's military business portfolio.

Production Rate Changes

MYP II followed immediately after the completion of the five full-rate production lots for MYP I. The production rate during MYP II was projected to hold steady near the high end of the annual rates achieved during MYP I, and at the same rate as the last lot of MYP I.

Program Maturity

F/A-18E/F MYP II is taking place during the mature phase of the production program. However, major planned ECPs and upgrades are taking place during this period, including the installation of the APG-69 Airborne Electronically Scanned Array fire control radar. It is unclear at this time how long production will continue beyond the end of MYP II. Current plans envision a rapid winding down of production after MYP II over three fiscal years.

Foreign Sales

On March 6, 2007, the Australian government confirmed that it intended to purchase 24 F/A-18Fs for the Royal Australian Air

Force. According to press reports, the Indian Air Force may purchase F/A-18E/Fs and has a requirement for up to 126 aircraft. There are other potential foreign sales prospects, but none has been confirmed.

References

Air Force Materiel Command, *Acquisition Reform Cost Savings and Cost Avoidance: A Compilation of Cost Savings and Cost Avoidance Resulting from Implementing Acquisition Reform Initiatives*, Wright-Patterson Air Force Base, Ohio, December 19, 1996.

Bodilly, Susan, Frank Camm, Richard Pei, *Analysis of Air Force Aircraft Multiyear Procurements with Implications for the B-2*, Santa Monica, Calif.: RAND, R-3990-DDR&E, 1991.

"C-17 Aircraft: Comments on Air Force Request for Approval of Multiyear Procurement Authority," testimony before the Senate Subcommittee on Seapower, Committee on Armed Services, March 28, 1996.

"C-130J Acquisition Program Restructured," *Defense Industry Daily*, November 9, 2006. As of April 17, 2007:
http://www.defenseindustrydaily.com/2006/11/c130j-acquisition-program-restructured/index.php

Congressional Research Service Report to Congress, *Military Aircraft: C-17 Cargo Aircraft Program*, IB93041, March 20, 2000, p. 3.

———, *Military Airlift: C-17 Program*, RL30685, May 30, 2006.

———, *Military Airlift: C-17 Aircraft Program*, RL30685, January 25, 2007.

Department of the Air Force, *Department of the Air Force Justification of Estimates . . . Aircraft Procurement, Air Force.* Documents published annually from 1980 to 2007 with slightly varying names. As of June 23, 2007:
http://www.dtic.mil
http://www.defenselink.mil/comptroller/defbudget

Department of Defense, News Release, Number 138-96, *Defense Acquisition Pilot Programs Forecast Cost/Schedule Savings of up to 50 Percent from Acquisition Reform*, March 15, 1996.

———, *E-2C Hawkeye*, MYP-1 Exhibit, Multiyear Procurement (MYP) Criteria, P-1 Shopping List, February 1998.

———, *F/A-18E/F (Fighter) Hornet*, Exhibit MYP-1, Multiyear Procurement Criteria, P-1 Shopping List, February 1999.

———, Instruction 5000.2—Operation of the Defense Acquisition System, Sec. R—Mandatory Procedures for Major Defense Acquisition Programs (MDAPS) and Major Automated Information System (MAIS) Acquisition Programs, Chapter 2—Acquisition Strategy, April 5, 2002.

———, *C-17*, Exhibit MYP-1, Multiyear Procurement Criteria, February 24, 2003a.

———, *CC-130J (USAF) and KC-130J (USMC),* Exhibit MYP-1, Multiyear Procurement Criteria, February 2003b.

———, *E-2C FY 04/05/06/07 Aircraft and Engine Procurement*, Exhibit MYP-1, Multiyear Procurement Criteria, P-1 Shopping List, February 2003c.

———, *F/A-18E/F (Strike Fighter) Hornet/EA-18G (Electronic Attack) Hornet*, MYP-1, Multiyear Procurement Criteria, P-1 Shopping List, February 2003d.

———, Instruction 5000.4-M-1—Contractor Cost Data Reporting (CCDR) Manual, February 2004.

———, Office of the Inspector General, *Contracting and Funding for the C-130J Aircraft Program*, D-2006-093, June 21, 2006.

Department of the Navy, *Department of the Navy Justification of Estimates . . . Aircraft Procurement, Navy.* Documents published annually from 1981 to 2007 with slightly varying names. As of June 23, 2007:
http://www.dtic.mil
http://www.defenselink.mil/comptroller/defbudget

Dews, Edmund, and Michael Rich, *Multiyear Contracting for the Production of Defense Systems: A Primer*, Santa Monica, Calif.: RAND, N-1804-AF, 1982.

DoD—*See* Department of Defense.

"Doing Business Differently" presented by the Boeing Production Pricing Group to RAND analysts, Boeing, Long Beach, Calif., October 7, 1999.

F-16 System Program Office, *F-16 Multi-Year Procurement Experience*, Memorandum, undated.

———, Validation of Multiyear Savings Associated with the Production of 720 F-16 Aircraft (FY86–FY89 Requirements), Contract F33657-84-C-0247, P00100, September 1986.

F-22 Program Office, "Appendix A–Program Office Estimate (POE06): Production Cost Estimate," given to RAND on November 29, 2006.

"F-22A Raptor: Division of Work," Lockheed Martin, homepage, 2005. As of April 17, 2007:
http://www.lockheedmartin.com/wms/findPage.do?dsp=fec&ci=15116&rsbci=15047&f

FAR—*See* Federal Acquisition Regulation.

Federal Acquisition Regulation Part 15—Obtaining Cost or Pricing Data, Sec. 403, Obtaining Cost or Pricing Data, September 28, 2006.

———, Part 17–Special Contracting Methods, Sec. 104, General, September 28, 2006.

———, Part 217—Special Contracting Methods, Sec. 103, Definitions, September 28, 2006.

———, Part 217—Special Contracting Methods, Sec. 172, Multiyear contracts for supplies, May 12, 2006.

Fisher, Gene H., *Cost Considerations in Systems Analysis,* Santa Monica, Calif.: RAND, R-490-ASD, 1970.

Foelber, Robert, *Cutting the High Cost of Weapons*, The Heritage Foundation, Backgrounder #172, March 16, 1982.

GAO—*See* General Accounting Office.

General Accounting Office, *Testimony of Walton H. Sheley, Jr., Director, Mission Analysis and Systems Acquisitions Division, to the Defense Task Force, House Committee on the Budget*, GAO-114658, March 10, 1981.

———, *Analysis of DoD's 1986 Multiyear Candidates*, GAO-86-1, November 1985.

———, *Procurement: An Assessment of the Air Force's F-16 Aircraft Multiyear Contract,* GAO/NSIAD-86-38, February 1986.

———, *DoD Procurement: Multiyear and Annual Contract Costs*, GAO/NSIAD-88-5, October 1987.

———, *Procurement: Multiyear Contracting and Its Impact on Investment Decisions*, GAO/NSIAD-88-125, May 1988.

———, *Assessment of DoD's Multiyear Candidates*, GAO/NSIAD-89-224BR, September 1989.

———, *Procurement: Assessment of DoD's Multiyear Contract Candidates for Fiscal Year 1991*, GAO/NSIAD-90-270BR, August 1990.

———, *Navy Aviation: AV-8B Harrier, Remanufacture Strategy Is Not the Most Cost-Effective Option*, GAO/NSIAD-96-49, February 1996.

———, *C-17 Aircraft: Comments on Air Force Request for Approval of Multiyear Procurement Authority*, Statement of Louis J. Rodrigues, Director, Defense Acquisition Issues, National Security and International Affairs Division, to the Subcommittee on Seapower, Committee On Armed Services, Senate, GAO/T-NSIAD-96-137, March 28, 1996.

———, *Options Exist for Meeting Requirements While Acquiring Fewer C-17s*, GAO/NSIAD-97-38, February 1997.

GlobalSecurity.org, "C-17 Globemaster III Production," April 27, 2005. As of April 17, 2007:
http://www.globalsecurity.org/military/systems/aircraft/c-17-prod.htm

Harshman, Richard A., "Multiyear Procurement in the First Reagan Defense Budget," *Program Manager*, September–October 1982.

Institute for Defense Analyses, *Acquiring Major Systems: Cost and Schedule Trends and Acquisition Initiative Effectiveness*, March 1989.

"Integrated Defense Systems: F-22 Raptor," Boeing Web site, 2007. As of April 17, 2007:
http://www.boeing.com/defense-space/military/f22/index.html

Lorell, Mark, and John C. Graser, *An Overview of Acquisition Reform Cost Savings Estimates*, Santa Monica, Calif.: RAND, MR-1329-AF, 2001. As of June 23, 2007:
http://www.rand.org/pubs/monograph_reports/MR1329/

Nelson, J. Richard, Bruce Harmon, Scot Arnold, James Myers, John Hiller, M. Michael Metcalf, Harold Balaban, and Harley Cloud, *F-22A Multiyear Procurement Business Case Analysis,* Alexandria, Va.: Institute for Defense Analyses Paper, P-4116, 2006.

Sparks, Col Gregg, *System Program Office Nineteenth Quarterly Reduction in Total Ownership Cost (RTOC) Forum*, briefing, August 4, 2004. As of June 23, 2007:
https://acc.dau.mil/CommunityBrowser.aspx?id=46577

USC—*See* United States Code.

United States Code, Title 10, Section 2306b, Multiyear contracts: acquisition of property, May 30, 3006.

Younossi, Obaid, David E. Stem, Mark A. Lorell, Frances M. Lussier, *Lessons Learned from the F/A-22 and F/A-18E/F Development Programs*, Santa Monica, Calif.: RAND Corporation, MG-276-AF, 2005. As of June 23, 2007:
http://www.rand.org/pubs/monographs/MG276/

Younossi, Obaid, Michael Kennedy, John C. Graser, *Military Airframe Costs: The Effects of Advanced Materials and Manufacturing Processes*, Santa Monica, Calif.: RAND Corporation, MR-1370-AF, 2001. As of June 22, 2007: http://www.rand.org/pubs/monograph_reports/MR1370/